approaches to ENVIRONMENTAL STUDIES

General Editor: Gordon A. Perry   book 23

Plant and Animal Habitats

in town and country

Gordon E. Simmons, F.L.S.

Senior Lecturer in Field Studies
Worcester College of Education

BLANDFORD PRESS

POOLE   DORSET

Acknowledgments

Ardea Photographics 3:9, 3:10, 3:15 and 4:12; and jointly with the photographers for 2:1 (I. R. Beames), 3:8 (Ake Lindau), 3:14 and 9:15 (J. B. & S. Bottomley), 4:8 and 7:23 (J. L. Mason), 9:3, 9:4, 9:6, 9:7, and 9:8 (John Clegg)
Bruce Coleman Ltd for 2:17 (E. Breeze-Jones), 7:1 (Malcolm Pendrill), 7:5 (John Markham)
M. J. D. Hirons for 3:5 and 5:15
Leslie Jackman for 8:9
Educational Productions Ltd for permission to use the author's photographs 4:7, 5:17, 7:2, 7:3, 7:10, 9:13 and 9:14, already used by them in filmstrip form.

All other photographs were provided by the author Gordon E. Simmons, who also drew the notebook 'sketches from life'. The other line illustrations were prepared by Helen Downton from the author's preliminary drawings.

First published 1976

Blandford Press Ltd

Link House, West Street, Poole, Dorset, BH15 1LL

© 1976 Gordon E. Simmons

ISBN 0 7137 3356 X

Printed and bound in Great Britain by Jarrold & Sons Ltd, Norwich

Contents

Editor's Note

This book is part of a wider scheme, *Approaches to Environmental Studies*, published by Blandford Press. This consists of class books, programmed texts, teachers' books and ancillary material. Programmes, involving the study of the environment and not requiring teaching machines, have been written by Colin Kefford. These are also available from Blandford Press.

Specially prepared illustrations in the form of 2 × 2 in. slides in colour to help in these studies may be purchased from: The Slide Centre, Ltd., 11 Bellevue Road, London SW17.

Teachers are recommended to use (*a*) the Handbook and (*b*) Guide No. 4 which supports and expands the ideas in the class books and introduces other areas of experiment and investigation.

G.A.P.

1-EVERYWHERE A HOME

It would be a big mistake to think that the countryside is the only place where you can study nature. Plants and animals of all kinds are to be found living in towns and cities, but these may not at first glance seem as exciting as those to be found living in such places as woodlands, meadows or streams. There may be no rare orchids growing amongst the pile of bricks on that piece of waste ground you pass on your way to your town school, but the many plants, often referred to as 'weeds', which grow there may be just as remarkable as the orchids. Many of these 'weed' plants have been introduced from countries far and wide but because they are living on our doorsteps we often pass them by with hardly a glance. Yet how much do we really know about these familiar plants and about the private lives of the animals associated with them? There is often as much to observe, even to cause surprise, in a few square metres of your own garden as in an acre of the wildest woodland, provided you keep your eyes open and really look.

This book is about plants and animals in their habitats. Mostly it is about those plants and animals to be found living in towns and cities, although we shall also explore the verges by our road-sides, the nearby common and the local canal.

What is a habitat?

The study of all forms of plant and animal life, together with the surroundings in which they live is called *Ecology*. Ecology was invented from the word 'οικος', Greek for house or home. Every time we plant a bush, a tree, some flowers, or even make a lawn, we are creating a home for the animals that will live there. A place where plants and animals live is called a *habitat* (home). A tree, a piece of waste ground, a hedge, a wall, a canal, are all examples of habitats. However, not all habitats are large. You may find plants growing in a crack between paving stones, or in the corner of a wall where dust and dead leaves have collected. You should remember that the habitat comprises the whole environment of the plant or animal and this includes the soil and various climatic factors to which the plant or animal is subjected. The sunny side of a wall may provide an environment quite different from the shady side.

1:1a A heap of soil left by the builders quickly becomes a plant jungle

1:1b Insect visitors to the garden runner beans include a Bumblebee and Ladybird Beetle

1:1c A dry stone wall in a school garden provides a more hostile environment

1:1d For a wetter habitat look at lockgates and walls of a canal

Plant and animal relationships

Whether a particular kind of creature will be able to live in any one special place will depend to a certain extent upon what other kinds of animals and plants are living there. Different kinds of plants tend to live in different places (habitats) and feed on different kinds of food, but all the plants and animals living in one place depend upon each other to some extent and form a *community*. However, wherever we find living things, and whatever their form or size, we will find that the habitat provides certain basic conditions for them. All living things need a place to live, air to breathe, water and food, in order to grow and reproduce their kind.

Food chains and webs

Habitats are worlds in miniature, but quite often very savage worlds, since life for many animals is one of eating and being eaten. An example is the world of a Rose bush in summer. Hordes of Greenfly or Aphids jostle each other as, with sharp mouth parts buried in the leaves and soft new shoots, they constantly feed from the ever-flowing sap. The excess sugary fluids given out at the hind ends of their fat bodies are eagerly collected by the Red Ants living in the lawn, whilst the aphids themselves are being devoured by the ever-hungry larvae of Ladybird Beetles, Lacewings and Hover-flies. Various small birds such as Blue Tits may feed upon those larvae as well as on the aphids and ants. This chain of events is called a *food chain* and may be represented by a series of arrows:

Rose bush → Aphids → Ladybird larvae → Blue Tit.

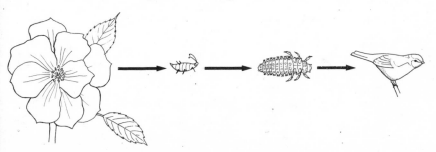

1:2   A simple Food Chain in the garden

However, in practice many animals feed upon more than one kind of food so there are other simple food chains we could draw, all starting from the green plant, the Rose bush and its flowers. These different food chains can be joined together to form what we can call a *food web*. Whenever plants are involved we should include the energy that all green plants obtain and utilise from the sun.

Green plants and decay

You can see how complicated is the food web we have produced from our Rose bush and its inhabitants (Fig. 1:3). In all food webs there are two important things to remember: the one concerning the sun and the green plant has already been mentioned; the other is that all living things die in the end and their remains become decomposed by bacteria and other tiny organisms. Thus the chemicals that they are composed of are released and can be used by other plants. When you are reading the chapters that follow see if you can write down the food chains and food webs that are illustrated there.

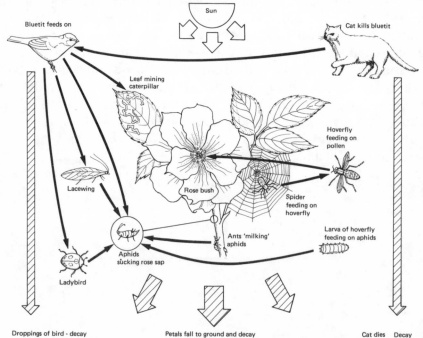

1:3   A Food Web on a rose bush. Which creatures are herbivores and which are carnivores?

Herbivores and carnivores

Another way of looking at food webs is to think of plants (such as our Rose bush) as being producers of food, and animals as being the consumers. The animals (such as aphids) which feed directly upon the plants we call *herbivores*. The animals, such as the larva of the Ladybird and the Blue Tit, which feed upon other animals we call *carnivores*. Both the herbivores and carnivores are consumers, but at different levels.

Effects of man

Sometimes man unwittingly upsets the delicate balance which occurs in these patterns of life in nature. This is particularly so in gardens and in farms. Perhaps he decides to kill the aphids on the roses by spraying them with a chemical insecticide, but as well as killing most of the aphids he also kills the predators, those carnivorous Ladybird larvae. Aphids can multiply at such an enormous rate that if only a few are left alive they will eventually increase rapidly, this time without any predators being present.

When and how to observe

The big advantage of studying the plant and animal life on your doorstep is that in most cases you can spend as long as you like in watching and recording it. It will be much more accessible throughout the seasons and, more important still, much less is known about the lives of animals and plants of town and city habitats than of habitats in the countryside. If you make regular observations throughout the seasons you may contribute to our knowledge of wildlife, an exciting possibility indeed. The author well remembers the fascination of studying creatures in the small garden of his boyhood, in an industrial town in Suffolk. The butterflies attracted to the Michaelmas Daisies, the ants and aphids on the Rose bush, the slugs and snails of the small rockery, and the activities of earthworms on the pocket-handkerchief-sized lawn, gave hours of pleasure. A favourite author in those days was, and is even still, J. Henri Fabre, affectionately called by his friends 'The Insect Man', found the delights of a life time, as well as adventure and fame, in observing the weedy pieces of waste ground near his home. He had no elaborate equipment and his best instruments, as Fabre himself said, were time and patience.

How to record

At first you may not see many creatures living in your back garden but if you go on exploring, like Fabre, and really use your eyes, you will find more and more. If you wish to search more thoroughly then a magnifying glass will be very useful. Almost any type is better than none, but the small folding pocket type which gives about ten times magnification is ideal.

Always carry a notebook with you. Note down the various plants you come across living in both town and country habitats. Include the size, shape and colours of the creatures associated with the plants and, where practical, their numbers. By making quick sketches of the plants in their habitats you will increase the amount of detail that you notice. Usually it is best to make notes on the spot, but sometimes you may be so engrossed in watching the behaviour of a particular animal that the recording has to be done as soon as possible afterwards (Fig. 1 : 4).

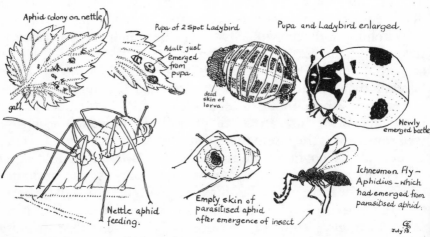

Aphid colony on nettle

gall.

Pupa of 2 Spot Ladybird

Adult just Emerged from pupa.

Pupa and Ladybird enlarged.

dead skin of larva.

Newly emerged beetle

Nettle aphid feeding.

Empty skin of parasitised aphid after emergence of insect

Ichneumon Fly – Aphidius – which had emerged from parasitised aphid.

July 78.

1:4 A page from the author's notebook

If you have a camera you may wish to take photographs of the habitats and the plants and animals living there. Photography, however, should supplement the notebook and not take its place. Imagine you are Gilbert White of Selborne, the country vicar whose observations in the eighteenth century made him one of the world's most famous naturalists. He had no camera but captured with his pen the most detailed observations of even the commonest animal and plant.

Discovery Assignment 1

1 Read the book *The Insect Man* by E. Doorly (Penguin) which describes the observations of Fabre. You should be able to borrow this from your library.
2 Visit your local museum and find out about the plants and animals which have been collected from your county.
3 Visit a library or local archive to look for old maps of your district. Ask permission to make copies of parts of maps which relate to your own study area. Make a list of major habitat changes that the maps reveal.

2-AROUND THE HOUSE AND GARDEN

Visitors in the house

Numerous small creatures leave their natural surroundings on occasions and enter our houses. Many, like spiders, Clothes Moths and Silverfishes, must have visited man's abode for centuries. Most people dislike and kill them; others just ignore them. You have an opportunity to study the lives and habits of these uninvited household guests. As well as listing the animals you find, ask yourself questions such as 'Why should this creature be in this part of the house?' You will soon realise that there is a pattern in their presence. The primitive shining Silverfish (*Lepisma saccharina*) (Fig. 2:1), which is covered with dust-like scales, is normally found in damp places such as beneath a box in the larder. Occasionally they are found amongst books, feeding upon the starch of the bindings. There too you may find the tiny pale-coloured Booklice (*Trogium pulsatorium*). These are quite similar to the lice found on birds and it is possible that they may originally have lived inside birds' nests. Clothes Moths are to be found in dry warm places, but it may be that modern central heating provides too dry an environment for them.

Outside the house and in the garden

Although no two gardens are exactly alike there are a number of common features all providing habitats for plants and animals. Many parts of your school surrounds may perhaps be similar to your garden. When you are reading Chapter 4 bear this in mind.

2:1 Look for Silverfish in damp places. Sometimes called the three-pronged Bristle-tail, it is about 8–10 mm long

2:2 How many different habitats for plants and animals are there in this garden?

A spider survey

What better creature to study in your garden than the spider? Indeed you will already have noticed the mat-shaped webs of the House Spider (*Tegenaria sp.*). All spiders have the ability to spin silk, but not all produce webs in which to trap their living food. Examples of those, the hunting spiders, will be mentioned in Chapter 8. Here we will consider the web spinners.

You will not have to go very far to find the untidy mesh webs of the Sieve and Comb Spider (*Ciniflo sp.*) (Fig. 2:3) for they will be across the corners of windows, on wooden fences and posts and on the brickwork above doorways. The threads of silk are whitish but quite often the new threads appear bluish in colour. Remains of the victims of this spider are often found on the web and inside its tubular retreat. These would make an interesting collection.

Exquisite orb-webs may be found across the corners of window frames and of course slung amongst the leaves and stems of tall plants and shrubs. Far commoner than the orb-web of the Garden Cross Spider (*Araneus diadematus*) will be those of the Telegraph Spider (*Zygiella sp.*) (Fig. 2:4). Notice that in the orb-webs of this species there are two segments missing. Whereas the Garden Spider is often to be found hanging upside-down in the centre of its large orb-web, more often than not the Telegraph Spider secretes itself away in a small hole, from which one of the radial silken spokes stretches, acting as a signal to the spider when anything alights.

2:3 *above, left* The untidy web of a Ciniflo Spider contains the remains of its last meal – a cranefly

2:4 Look at the above picture for the missing segments in the orb-web of the Telegraph Spider

2:5 A hammock web made by a tiny Money Spider

Amongst the low-growing shrubs and plants will be the hammock webs of the Money Spiders (*Linyphiid triangularis*) with their maze of irregular scaffold threads above (Fig. 2:5). A group of smallish, round-bodied spiders (*Theridion sp.*) spin a number of leaves together in a simple web. Look for these on Lilacs, Flowering Currants and other soft-leaved shrubs.

Cultivated plants and weeds
Most people's interest in weeds is a consuming desire to get rid of them. Do get to know the weeds of your garden, for many will appear when you make surveys along the street, in the school grounds and on waste land. Others, however, may only be found in your garden, for each type of soil has its own special weeds. A weed, of course, is any plant growing where we do not want it. They are so successful because they are able to produce immense quantities of seed in relation to their size. For instance Groundsel (*Senecio vulgaris*) may produce something in the region of 1,000 seeds per plant, whilst Shepherd's Purse (*Capsella bursa-pastoris*) produces 4,000 or so seeds. Find out the appropriate figure for a plant of Great Plantain (*Plantago majus*). Weed species may also differ according to the soil cultivation. Compare those found on a path with those beneath shrubs and with an annual border. How do these weeds arrive?

2:6a *left* Rosebay Willow herb
2:6b Its plumed fruits are released in enormous numbers

Individual plant studies

'Adopt' a particular plant and find out as much as possible about it. A most prolific weed is Rose-bay Willow Herb (*Chamaenerion angustifolium*) (Fig. 2:6) yet about 170 years ago it was very scarce in this country. People living in London were surprised to find this plant, called 'Fireweed' by Americans, together with Oxford Ragwort (Fig. 5:5, page 43) amongst the first plants to brighten the scenes of destruction in the bombed areas during the war. Seed production is staggering, 80,000 plumed seeds being produced by the average plant. It is also a perennial and is the larval food plant of the attractive Elephant Hawk Moth (Fig. 2:7). The flowers produce a bright blue pollen much sought after by bees.

Flowers and their insect visitors

Many flowers depend on insects to reproduce themselves. Insects normally only visit flowers to collect pollen or nectar, the fertilisation of the flowers and resulting seed production being a by-product of their activities. What can you find out about the insects that visit your own garden flowers? Although some insects arrive 'on foot' (e.g. worker ants), most insects fly from flower to flower, including a number of beetles (Fig. 2:8).

2:7 Larva of Elephant Hawk Moth feeding on willow herb

2:8 Sailor Beetles mating on cow parsley flowers

An insect can only reach the nectar of a flower if its tongue or 'proboscis' is long enough. Most beetles and flies, such as the Hover-fly (Fig. 2:9) have short prosboces, whereas many bees and almost all butterflies have long prosboces. Some insects, like wasps and ants bite through the flower to get to the nectar, so they do not aid the plant in pollination. Watch out for creatures that lurk inside flowers waiting to pounce upon insect visitors as they arrive.

Butterfly bushes and flowers
Some flowers, by secreting more nectar than others, are particularly attractive to butterflies. Buddleia (Fig. 2:10) is a good example. Originally introduced to this country from China, Buddleia (*Buddleia davidii*) has colonised a number of waste places in towns (Chapter 5). The long flower spikes smell strongly of honey. In August and early September you should see such butterfly visitors as Small and Large Whites, Small Tortoiseshell, Peacock, Comma and Red Admiral.

A number of herbaceous and annual flowers also attract butterflies. Look for the pink strain of Valerian and the blue-flowered Catmint, which is often visited by the Green-Veined White Butterfly (Fig. 2:12). The old-fashioned variety of Golden Rod (*Solidago sp.*), the later flowering flat-headed pink *Sedum spectabile* or Ice Plant and Michaelmas Daisies are often crowded with butterflies jostling to get their fill of nectar.

Record the feeding habits of butterflies, for some have the quaintest of tastes. The Red Admiral Butterfly (Fig. 2:11d) for instance 'adores' the liquor from fallen apples and is also partial to newly laid tar on roads!

2:9 A Hover-fly (Syrphid species) rest on a garden rose in autumn

2:10 Peacock Butterfly feeding on Buddleia

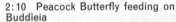

a   Large White, on flowers of Phlox

b   Small Tortoiseshell

c   Comma in autumn

d   Red Admiral

2:11   Butterflies to be seen in the garden.

2:12   Green-Veined White Butterfly on Dandelion

The open lighted window

Many insects are attracted to light, particularly on warm summer evenings. The author has recorded moths, beetles, Lacewings, Crane-flies, Ichneumon Flies and a great variety of two winged flies and midges coming to his bathroom light. After obtaining permission, try leaving one of your lights on for an hour or so and collect all the creatures that fly through the open window. Which are the commonest groups to arrive? Compare the catches on other nights under different weather conditions.

Life by torchlight

A walk around the garden after dark, periodically using a torch, will reveal the nocturnal feeders and their hunters. Look at flowers of Honeysuckle, or of the Tobacco Plant (*Nicotiana*) for these have long-tongued moth visitors. Slugs and snails will be slowly moving amongst flowers, through grass, or over decaying material on the compost heap. Collect specimens of the different kinds you find, putting them into separate containers with a note of what they were feeding. They are not difficult to identify. It is surprising how few of our slugs actually damage garden plants.

Look upwards and you may see tiny Pipistrelle Bats hawking low to catch their quota of night-flying insects. Shine your torch into the Rose bush and you may spot the golden eyes of the Green Lacewing (*Chrysopa sp.*) as it feeds upon the hordes of aphids.

Aphids, Ants, Ladybirds and Lacewings

Since aphids are so widespread you should be able to use your observations to answer the following questions:

1 Which plants are infested with aphids?
2 Are all aphids the same colour?
3 How and when do they arrive on a particular plant?
4 Have they all got wings?
5 Do any aphids die – are they parasitised by other creatures?
6 What is 'honey-dew'? What is the association between ants and aphids? (See pages 7 and 48.)
7 What are the predators of aphids? (See Figs. 1:3 and 2:13.)

Wasp studies – hibernation

It is the queen wasp which hibernates. All the other workers and males die at the onset of winter. The queen wasp, already fertilised by a male during the previous autumn, emerges in spring and quickly finds a place to build her nest. Using her jaws as chisels she scrapes slivers of wood from perhaps the door of a shed and fashions a nest of a few cells (Fig. 2:15) suspended from the roof of her nest hole. Into each cell she lays an egg which will eventually become a wasp worker. The wasp workers then take over the task of feeding the larvae, leaving the queen to con-

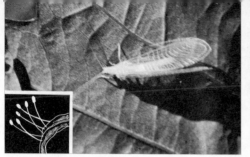

2:13 An aphid predator – Green Lacewing (*inset* stalked eggs)

2:14 Nocturnal feeders – Banded Snail on Iris leaf

centrate on egg laying. You might come across the new papery nest of the German Wasp (*Vespa germanica*) hanging from the roof of a garden shed or outhouse. The wasps feed the larvae with a paste made from chewed up insects, receiving in exchange a sweet fluid exuded by the grubs. It is not until the queen has stopped laying eggs in August that the inmates of the wasp city seek out other sources of sweet substances (Fig. 2:16). The grubs of a two-winged Drone-fly (*Volucella pellucens*) live in association with the wasps, acting as first-class scavengers for they feed upon the excreta of both wasps and wasp grubs.

By far the most species of wasps are not social but lead solitary lives. More will be said about these in Chapter 6.

House-martins – migration

House-martins, Swallows and Swifts migrate from Africa, where they have spent the winter, and arrive in Britain during late spring. They are summer residents only, leaving again in August or September for the warmer countries where their insect food is abundant. Some creatures hibernate when food is scarce, but many birds migrate. Other summer residents include the Nightingale, Spotted Flycatcher, Chiff-chaff, and Cuckoo.

Keep a time and weather chart showing both the arrival and departure dates of these summer visitors. The House-martin is a good bird to study. Although early arrivals may be seen in April the main mass arrive in the first half of May, when they return with remarkable navigational accuracy to the vicinity where they were born, and quite often to the same nest site.

Here are some observations and recordings which you can compare with those of your friends:

1 Date and time of arrival of the birds.
2 Commencement and method of nest construction.

3 Length of time for nest construction.
4 Where does the mud come from?
5 What part of the house or building is used by the martins?
6 What direction does the nest face? Is there a 'favourite' direction?
7 What is the colour of the building surface (a) behind the nest and (b) above the nest?
8 What happens if a House Sparrow attempts to take over the nest?
9 What happens at the end of the season?

Garden birds throughout the year

Most birds you see in your garden, such as Blackbirds, Thrushes and Robins are residents throughout the year. The best way of observing the birds of your garden is to put food out for them regularly. You will then soon attract many other species of birds even though you may live in a very built-up area. Do remember that in winter time they will become dependent on your generosity. If you forget to put out food on a hard, cold day, the birds may sit around for hours before they finally seek food elsewhere. Remember also that not all birds will come to a bird table; some, like the Blackbird, Thrush and Dunnock (sometimes called the Hedge Sparrow) will only feed upon the ground.

Water is a necessity in winter but should be available throughout the year. Although birds do not sweat they lose water, mainly by excretion, and must make up this loss. Some water will come from food but the rest will come from drinking. Water is also used in helping to keep their feathers in good condition. A bowl of water sunk into the ground will enable you to witness the complicated sequences of bathing and preening (Fig. 3:9, page 28).

2:15 Wasp nest made by the queen in a gooseberry bush during early May

2:16 Worker wasps feeding on apple in autumn

2:17 A Great Tit helps itself to the cream from a milk bottle

Discovery Assignment 2

GARDEN BIRD PROJECTS

1 Bird lists. Make daily or weekly observations for the same period and at about the same time:

Birds seen 0820–0830 W/E 3 March 1974							
Day	Blackbird	House Sparrow	Song Thrush	Robin	Starling	Blue Tit	Notes
Monday	1	8	1	3	6	3	*Blackbird
Tuesday	2*	6	–	2	4	2	with white
Wednesday	1	4	–	2	4	3	tail feather
Thursday	2*	2	1	1	8	1	chased
Friday	1	8	1	3	9	1	away by a
Saturday	1	8	2	2	6	–	normal-
Sunday	2	5	–	2	5	2	coloured
							bird
Weekly total	10	41	5	15	42	12	
Daily average	1	6	1	2	6	2	

2 Find out about bird feeding habits:

(*a*) The types of food eaten by different bird species.

(*b*) The times of feeding. Do birds eat constantly throughout the day, or do they appear to have specified 'meal times'?

(*c*) Dominance: which bird species chase away others from the feeding area? Is there a 'safe' distance where a chased bird will be able to feed? (The species of Tits are good examples to watch.)

(*d*) Methods of feeding: how are apples eaten? Which species pick meat off suspended bones? How quickly does the House Sparrow act in the same manner as say a Blue Tit?

3-IN THE STREET

Have you ever thought how long the streets and roads so familiar to you have been part of the town or city as you know it? Perhaps the actual name of the street or road may give you a clue. If you are in a new housing estate, perhaps not so long ago the land was part of the open countryside, but if you live in an older area in London or some other large city you may have to go further back in history to discover its origin. You may even be able to discover old prints in a library illustrating the street before it was made up with asphalt or tarmac. Thus, even the street as a habitat is changing, nevertheless Nature is rarely completely absent, as you will discover if you look hard.

Trees, pavements, gardens and churchyards

Tree surveys
Some trees which you find growing in streets must have been there long before the pavements were put down. Lime (*Tilia vulgaris*), London Plane (*Platanus acerifolia*) and Sycamore (*Acer pseudoplatanus*) are likely to be the most common, but your survey may be different. Look at the results of one made in London a short while ago:

Locality : Church Road, Tooting			
Tree Species	*Numbers*		*Total*
Lime	̶I̶I̶I̶I̶		5
Elder	I		1
Sycamore	̶I̶I̶I̶I̶	I	6
Elm (alive)	̶I̶I̶I̶I̶		5
Elm (dead)	̶I̶I̶I̶I̶	IIII	9
Oak	II		2
Hawthorn	I		1
Laburnum	I		1
Evergreen Oak	I		1
Horse Chestnut	̶I̶I̶I̶I̶ ̶I̶I̶I̶I̶	II	12
Acacia	̶I̶I̶I̶I̶ II		7
	Total living trees		41
	Total dead trees		9

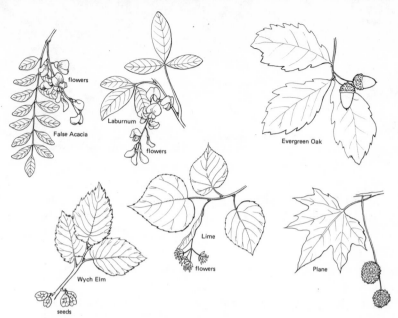

3:1 How well do you know your street trees?

As you might expect from the name of the road surveyed, a number of trees were found growing in the surrounds of a church. Strangely enough there was not an example of the Yew (*Taxus baccata*) which, because it is poisonous to man and other animals, was associated with mystical powers and with death.

The Evergreen Oak (*Quercus ilex*) was introduced from Mediterranean regions in the sixteenth century into churchyards and large gardens. The presence of a large tree in a new housing estate may indicate that the area may once have been part of a large private park. Elms, too, are seldom planted along streets because the twigs and branches, being brittle, often break off. The dead Elm trees, of course, had been killed by Dutch Elm disease, a fungus introduced by a species of Bark Beetle (Fig. 3:3).

How many of the trees illustrated in Fig. 3:1 have you seen in your own town streets? Both the Common Lime and London Plane trees are commonly encountered in most parts of London. Many of these must have been planted during the middle of the seventeenth century, or shortly afterwards. The Lime trees which once grew in Berlin gave the name to the famous 'Unter den Linden'. 'Linden' is an old word for 'lime'. The leaves, as well as providing a pleasant shade in summer, are a habitat for

3:2 Lime trees in a London street

3:3 Breeding galleries of the Elm Bark-beetle

aphids, which occur in such numbers that the leaves literally drip 'honey dew'. This sticky substance secreted by the aphids often coats the pavements below the trees. The sweetly scented flowers of the Limes which appear in late July are visited from morning until night by industrious Honey Bees because of their great nectar flow.

The London Plane is the most hardy of all city trees. It is the result of a cross between the Oriental Plane and the Western Plane. Perhaps its hardiness is due in part to the strange habit of its trunk, in that the bark peels off in large flakes, getting rid of the grimy outer layers. Those able to visit London should go and see the majestic Plane avenues along the Thames embankment and along the Mall from Admiralty Arch to Buckingham Palace.

Leaf feeding insects

Even in the city the foliage of many pavement trees will provide a habitat for a number of insects, including the larvae of moths. Of course much will depend upon the state of the atmospheric pollution, which could be tested (see Book 19, *Pollution and Life* by Arnold Darlington, in this series). The larvae of the Lime Hawk Moth and Peppered Moth may be found on Lime (Figs. 3:4), the hairy larvae of the Vapourer Moth on Plane and those of the Puss Moth and Poplar Hawk Moth on various Poplars, all growing in street pavements. If you stand beneath the canopy and look towards the sky many of the caterpillars present will show up as silhouettes against the thin spring leaves. Sometimes it is possible to find them on the pavements below trees after a sharp shower of rain, or in autumn as they search for a place to pupate.

3:4a Larva of Peppered Moth

3:4c Larva of Lime Hawk Moth

3:4b Adult Peppered Moth

3:4d Adult Lime Hawk Moth

Man and trees

What happens to the leaves which fall from the pavement trees in autumn? In most parks they are swept up into heaps. Outside the town or city there exists a natural recycling of the spent leaves of autumn, of fallen twigs and branches, but this does not happen in the town.

Autumn is a good time to make a collection of leaves. Look for evidences of animal damage. Some leaves may have small transparent blotches in them, others may have transparent winding galleries looking like pencil scribbles on the surfaces. These will have been caused by leaf mining insects – usually the larvae of tiny moths or flies.

Notice, too, that street trees often have peculiar shapes quite different from the same species growing in parks or open countryside. This is because their growth has been regulated by pruning, a practice called 'pollarding'.

Fences and garden hedges

Hedges provide some of the most important habitats in our countryside (Chapter 7) but hedges are also used to enclose gardens. Much will depend upon the type of hedge shrub, but even so each hedge will provide a possible habitat for all kinds of small creatures that are well worth studying. What is the commonest house boundary in your area? How do they compare with the results shown below?

Boundary Type	Church Road, Tooting, London No. of houses	Part of Oldbury Road, Worcester No. of houses
Wooden fence	14	2
Low brick wall with metal fence	12	–
Hedge of Green Privet	46	6
Hedge of Golden Privet	2	10
Hedge of Lonicera (Japanese Honeysuckle)	1	2
Hedge of Fuchsia	1	–
Hedge of Potentilla (newly planted)	1	–
Hedge of Lilac	1	2
Hedge of Box	–	2
Hedge of Hawthorn/Mahonia/Holly	–	2

The Lilac hedges were especially interesting for a number of the leaves had brown patches. On cutting one open there were four tiny whitish larvae living inside. Inside another patch there was nothing apart from some black pellets of excrement. However, some of the Lilac leaves had been neatly rolled down and fastened by little silken strands to the remaining leaf blade. Inside each roll was a larva, similar to those in the brown patches

3:5 Blotch mine in Lilac leaf caused by a moth larva (*Gracillana syringella*)

3:6 Edge of pavement colonised by Greater Plantain, Dandelion, grass and Pearlwort

(blister mines) but larger. When the larvae become too large for the blister mines they eat their way out and live inside the curled-up leaf, to pupate in the ground when fully grown. Eventually a minute but very beautiful moth emerges to carry on the life cycle.

Privet hedges are likely to be colonised by the larvae of the Small Ermine Moth, and also the Magpie Moth.

Cracks between paving stones

Even the narrowest chinks between paving slabs may become colonised by a wide variety of plants, from the lowly mosses to various grasses and other flowering plants. You can record these by making a simple list, by making a count along a particular stretch of pavement, or by making a sketch of the paving stones showing the distribution of the plants. However, in all cases you should look up in books how the various plant seeds might have arrived at the site being studied. Note down any evidences of animals such as the trails of slugs or snails, fine soil particles thrown up by ants, or aggregations of worm casts. Although many plants will have wind-dispersed seeds, a number may have been brought to the site by birds, or attached to the coats of animals such as dogs.

Bird inhabitants of town and city

In the built-up areas of town and city, such birds as the House Sparrow, the Starling and the Feral Pigeon scratch a living,

3:7 Greater Bindweed climbs anti-clockwise up iron railings

3:8 Starlings blacken the sky as they mass over their night-time roosting site

utilising almost any site amongst the tall buildings as a desirable residence to shelter and breed. After reading about these birds why not go out and learn a little more about the private lives of the birds of your streets?

The Starling (*Sturnus vulgaris*) is probably the commonest and most successful bird in the world, even surpassing the House Sparrow, for it is greedy, pushing and quarrelsome. Although it is now so common, yet 200 years ago Gilbert White, the Hampshire curate wrote of the Starling in one of his classic letters 'no number is known to breed in these parts'. Although the natural nesting site of the Starling is a hole in a tree, it will use any holes, cracks or openings in buildings, particularly below the roofs. Some Starlings stay near to their nest site throughout the year but the majority disperse across the country in flocks of a few hundred or so. By nightfall the flocks assemble to specific roosts, some in woods in the countryside but more often in vast dormitories in towns. Tracking down Starling roosts is one of the most absorbing forms of bird watching. To see the late afternoon winter sky darken with hordes of Starlings as they circle around and drop on to city trees and buildings, like a blizzard of black snow, gives one a great thrill. It has been estimated that some 100,000 Starlings daily commute to the buildings of London, fouling the ledges and pavements with their droppings. Roosts occur in various other cities but Glasgow is said to have the largest with at least 250,000 birds!

The House Sparrow Unlike the opportunist Starling the House Sparrow (*Passer domesticus*) is undoubtedly very dependent upon

3:9 House Sparrows 'enjoying' a communal bath

3:10 A gang of House Sparrows reveling in a dust bath

3:11 A London Sparrow overcomes its natural wariness of humans

3:12 London's Feral Pigeons feeding, preening and sunning themselves.

man. They seem to have a special talent for being always on the spot whenever there is easy food to be had. Have you ever noticed how Sparrows are the first to arrive when you scatter a few crumbs in a park, garden or street? Whereas birds like Robins, Blackbirds and Thrushes have specified territories during the nesting season this is not so with House Sparrows. They always seem to be in little gangs. It is because they are social birds that they are continually calling to each other.

As well as feeding in gangs, House Sparrows seem to enjoy other group activities such as bathing in the dust and in a puddle of water. Watch how they spread their wings out with bodies close to the ground, almost as if they are sunbathing! As you have already read, this is important for the care of the feathers.

House Sparrows are amongst the few birds in Britain that use their nests throughout the whole year. Whereas other birds

3:13 House Sparrows among the more exotic birds of St James's Park, London

roost on boughs of trees, Sparrows often sleep in their nests. Even in winter they can be seen taking wool or feathers into the nest, which has the unusual feature of being completely covered over. Sparrows seem to favour nest sites in buildings, preferably those that are occupied. Although very wary birds, under some circumstances they can become very tame (Figs. 3:11 and 3:13).

Look at the way Sparrows fly. Compare their flight pattern with that of the Starling, Pigeon and other birds. Watch them at the bird table or in a park catching flies or ants in flight. Look for partly albino birds – are these allowed to be one of the gang?

The Town Pigeon If there is a large flock of Pigeons in your town, roosting on the sides of buildings, similar to the 'London' Pigeons (Fig. 3:12) these will be Feral Pigeons. Feral Pigeons are descendants of the true Rock Dove (*Columba livia*). Rock Doves were kept in dovecotes during the Middle Ages but many escaped to live semi-wild around buildings.

The Collared Dove (*Streptopelia decaocto*) is about the size of an ordinary Feral Pigeon, although it appears much slimmer. There is a black line, or 'collar' half-way round its neck. Look closely at Fig. 3:14 and notice that the black collar has narrow white edges. The Feral Pigeon is very reluctant to use trees for perching whereas the Collared Dove may often be seen perching in trees, making its unmistakable song 'coo, cooo, cuk'! There may be Wood Pigeons in the nearby park but these have no collar and their call sounds like 'cooooo-coo, coo-coo, coo'!

The Collared Dove has a fascinating success story. This bird was never seen in the wild in Britain before 1952! Around 1920 the Collared Doves of the Middle East and India began to increase in numbers and move to places where they had not been seen before. Their rapid advance across Europe and Britain looks almost like a well-planned military campaign. From a small stronghold in the Balkans in the 1930s the birds had advanced to Holland by 1947, to Denmark by 1948, followed by Sweden in 1949 and Belgium in 1952 – when they were first seen in Britain. The first nest site was not recorded until 1957 and since then they have spread to most corners of the British Isles, with the population rising fast every year. Can you find out when they reached where you live?

3:14 Look out for the Collared Dove

3:15 Untidy nest of a Wood Pigeon in the scaffolding of a city building site

Town birds and you

Why do you think the bird population in Britain is gradually changing? For just as some birds are increasing so other sorts of birds are getting scarcer. This is where your observations around your town, if carried on for a number of years, might provide interesting information.

Discovery Assignment 3

1 On a large-scale street map plot the different trees growing there.
2 Write a simple booklet about the trees, which could be used as a 'self guiding' aid around your town 'tree trail'.
3 Make a survey of the hedges in your neighbourhood. Can you relate the type of hedges and their height to the age of the house, or of the people who live in the house?
4 Map the plants growing between paving stones, or at the bases of trees in pavements. If possible collect seeds from the plants, or similar ones, and sow them in pots, recording how well they germinate.
5 Plot the positions of as many nests as possible on or in the buildings along any street near you.

4-THE SCHOOL SURROUNDS

How familiar are you with the immediate surroundings of your own school? Could you write down the different habitats there? As each garden at home may be different so will many school estates. However, many habitats are common to all.

Looking up at roofs and gutters

Some old stone tiled roofs may have large colonies of algae, mosses and lichens. Compare the roofs of your school with those of other buildings in the neighbourhood. Do the pitch of the roof and the orientation of the building (direction in which the building faces) affect the plant cover? Old schools in the country-side may even be colonised by the House Leek (*Sempervivum tectorum*). Even if your school is comparatively new, if asbestos roofing tiles have been used these may have been colonised by lichens.

Some gutters often become clogged with silt and dead leaves. Quite often seedling plants of Groundsel, Willow Herb and

4:2 Colonisation of asbestos roof by Grey Cushion Moss (*Grimmia pulvinata*)

4:1a The lawn, path and hedge are but some habitats here. What others can you see?

4:1b A simple feeding table will bring birds even to a city school

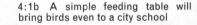

4:4 Silver Thread Moss (*Bryum argenteum*) on concrete steps. The pale patches in the foreground are lichens

4:3 A leaking rainwater pipe provides a habitat for algae, mosses and ferns

Sycamore may appear. Sometimes a gutter becomes so blocked that the rain water overflows and runs down a wall. In such areas there may be colonisation by algae (Fig. 4:3) and sometimes even ferns may appear.

Looking down at the car park

On this inhospitable habitat you will have to look closely. Much will depend upon the surface – whether it is smooth or rough, damp or dry, level or sloping. You will probably see just a silvery-green haze over the surface, but in some corners there may be tiny tussocks of a moss called the Silver Thread Moss (*Bryum argenteum*). Unlike most other mosses, the Silver Thread Moss does not often produce spore capsules. The silvery tips to the leaves are easily broken off and often root. (See *Pollution and Life*.)

Looking down at concrete paths

Much will depend upon the state of the path – whether it has any cracks or is made of separate paving stones. Is there any soil between the cracks? Are there any plants growing in the cracks? Compare your lists of plants with those made in the streets and at home. Is the distribution of the plants affected by where people habitually walk? In one particular survey the author made it was very noticeable that the Plantains, Pearlworts and Dandelions grew at the extreme edge of the pavement where it met the kerbstones. Compare the flora of a concrete path with one of hard packed earth, gravel or cinders. The shelter of steps may affect the flora too.

Beneath the paving stones all kinds of creatures may hide, such as snails, slugs, small beetles and woodlice. Worms very often throw up soil casts in cracks, and during the summer months on light soil you may see the activities of ants (Fig. 4:5).

The flower bed and shrubberies

Plant life

Are the weed plants growing in say a rose bed and a well established shrubbery similar? Have you ever wondered where all these plants come from? Seeds can remain dormant in the soil for a long time, in some cases many years, waiting for the right conditions to allow them to germinate (see page 45). Is it that the cultivations are different or are there more dormant seeds in the soil of each bed? You could test this by putting samples of soil into separate shallow boxes (such as those obtainable from a fishmonger's shop). Remove any large stones and pieces of root and firm the soil down, particularly at the edges of the box. If dry, water but avoid making it waterlogged. Put each box into a polythene bag with a label and keep in a warm place. When the first plants begin to appear remove the polythene, making a weekly plan on squared paper of each box.

You may be unfortunate in not having any such beds at your school. Never mind – you could carry out this project. Obtain some sterilised soil from a gardening shop and fill a couple of shallow wooden boxes. Water well and put them into your playground. One box could be on the ground and the other on a wall.

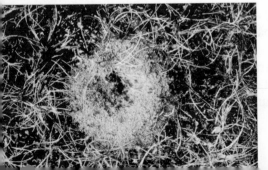

4:5 Black Ant workers emerging from their nest below a sandy path

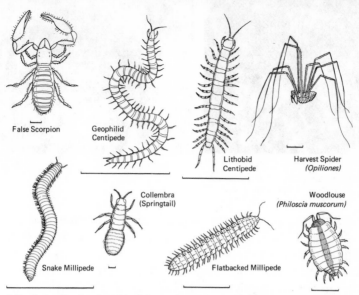

4:6 Animals of soil and leaf litter – lines show relative sizes

Plot any plants which occur over a number of months. These plants must have arrived as seeds, or fragments of plants (propagules) after the boxes had been put out.

Animal life

You may have watched certain birds pecking amongst the fallen leaves below a shrub. What might they have been searching for? Collect some leaf litter and the top few centimetres of soil from below a shrub, put it on a large sheet of white paper in the classroom and sort through it for animal life. Put the animals, many of which will be small, into separate containers. Fig. 4:6 shows a selection of animals you are likely to find. Springtails (*Collembola*) and Millipedes are both herbivores. Both the types of Centipede shown, however, are carnivores. Many of the soil inhabitants are even smaller. To extract these you will have to fit up a Tullgren funnel, which is explained on page 22 of *Pollution and Life*.

The rockery

Many invertebrate animals have such thin skins that they need protection from the sun's rays, which would quickly reduce the moisture in their bodies. A rockery, or even a pile of stones,

4:7 Two species of woodlouse. The Common Garden Woodlouse (*Oniscus asellus*) and the darker and rougher *Porcellio scaber*

4:8 Unlike most woodlice, the Pill Woodlouse can curl into a ball

forms a habitat for such creatures. After getting permission, carefully lift one of the smaller rockery stones to find what is sheltering beneath. Record the kind of stone – whether it is of limestone, sandstone, granite, or perhaps broken concrete. Do the animals seem to go for rough or smooth surfaces? The amount of available lime, which is often present in builders' mortar, affects the kind of small animals present. One of the invertebrates likely to be encountered is the Woodlouse (Fig. 4:7). These feed mainly on decaying plant material, which after passing through their bodies appears as tiny pellets of soil-like material. The cuticle (skin) of the Woodlouse is hardened with chalky material similar to that in a snail shell, much of which is obtained from the food. However some species, like the Pill Woodlouse (*Armadillidium vulgare*) (Fig. 4:8) has a much thicker cuticle. It is therefore found in habitats where there is a good supply of the chalky substance (calcium). Because the cuticle is thicker than other species it is also found in drier habitats.

There are various species of Woodlouse which have become adapted to widely differing habitat conditions. See how many you can find in your school grounds. *Remember, however, always to replace a stone carefully in the same position as it was found.* You can quite quickly destroy a habitat and the creatures living there by carelessness and also by too much interference.

Lawns and playing fields

Do we take grass for granted? We should not, because grass provides man, either directly or indirectly, with his food. Let us have a closer look at the grassy habitats to be found near schools.

36

4:9a Rosette weeds of Dandelions, Daisies and Plantains on a lawn

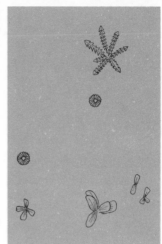

4:9b The same represented as a line map

Lawn weeds

Many gardeners and groundsmen spend much time trying to get rid of the various broadleaved plants growing in competition with the associated grass species. Walk over your lawn or playing field and carefully collect specimens of the weed plants you see amongst the grass. Most will be growing in a rosette fashion. Compare their root depths and types of root storage organs with grass roots. Do these lawn plants ever manage to produce flowers and seeds? (See also pages 73 and 74.)

Earthworms

Autumn is probably the best time to investigate the earthworm populations of playing fields and lawns. Look for the worm casts on the surface. There are two main types of earthworms. *Allolobophora* excavates galleries below ground and excretes fine soil in little 'casts' on the surface after extracting the humus particles in its digestive system. *Lumbricus*, however, deposits its casts way below the surface but uses a variety of material to plug the ends of its burrows. You may find small pebbles piled like little cairns over the entrances, but more often various leaves, and even sycamore seeds, pulled into its burrows. A famous scientist, Charles Darwin, once calculated that earthworms can bring up to the surface of the ground over 25 tons of soil per acre each year in the form of 'casts'.

4:10 Solitary Mining Bee emerging from its breeding gallery

4:11 A swarm of Honeybees collecting on a tree branch in the playground of Henwick Grove Primary School, Worcester

Solitary Mining Bees

If your lawn is on light sandy soil your sharp eyes may spot tiny pyramids of fine soil appearing in spring. Although at first glance they might appear to be the result of the ants' spring-cleaning activities they are more likely to be the work of a species of Solitary Mining Bee, possibly *Andrena fulva* (Fig. 4:10). The newly emerged female bees sink shafts into the ground, constructing a special earthen cell into which an egg will be laid, after they have first stocked this cell with balls of pollen and nectar, a kind of 'bee bread'. Each solitary bee works entirely on her own, solely for her own future progeny, even though there may be a number of other females constructing their own burrows in close proximity.

The school 'short cut'

What happens to the grass on those areas where you and your fellow pupils take short cuts? Even animals such as dogs and cats take short cuts across areas of open ground. If these are used regularly, after a while they begin to show. At first the area becomes compacted and hard, then the grass becomes thin and worn, until finally it may become completely bare of grass, apart from a few plants that can withstand treading. This would be a good area in which to record the recolonisation of a bared area. Put up a sign, 'Do not use this short-cut! Conservation area!' The pressure of human feet can soon alter the vegetation of an area.

Nests and nest boxes

Even in a noisy school birds will nest in sheltered parts of the buildings. It is surprising where birds in towns and cities will nest and breed. The following are a few examples. A Robin nested in a filing tray by flying through the open window of a study, whilst another constructed its nest between the saddlebag and seat of a bicycle. A House Sparrow nested in an old hen coop, a Hedge Sparrow in a small Lavender bush, and a Blackbird successfully reared her brood on top of a tray of Tomato seedlings inside a greenhouse! However exciting it is to discover where a bird is nesting, do remember that the welfare of the bird and its young must come first. Do not disturb them.

As well as leaving an old coat, a pile of old pea sticks or a heap of rose prunings by a shed you may wish to provide more artificial nesting sites in the form of bird nesting boxes. There are two basic types as you can see in Fig. 4:13, one suitable for various Tit species, whereas the 'open-plan' type is better for Robins and Blackbirds. Ideally the boxes should be put up as early as possible during the winter, but a box put up in March was within a few hours occupied by a pair of Great Tits. This, however, is probably an exception.

4:12 Nesting sites in towns are often different from those in the agricultural countryside. Here a Robin nests inside an old kettle

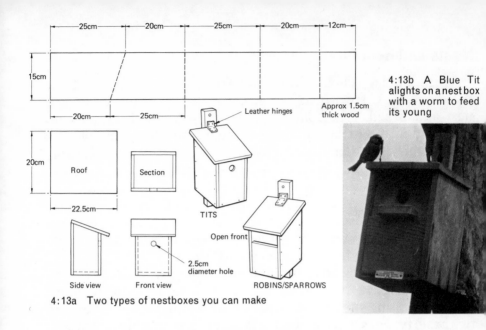

15cm

25cm | 20cm | 25cm | 20cm | 12cm

20cm | 25cm

Approx 1.5cm thick wood

Leather hinges

20cm

22.5cm

Roof

Section

TITS

Open front

2.5cm diameter hole

Side view | Front view

ROBINS/SPARROWS

4:13a   Two types of nestboxes you can make

4:13b   A Blue Tit alights on a nest box with a worm to feed its young

Where can you fix them? Most hole nesting birds seek sites anywhere between 1·5 and 6 m above the ground. *Don't* put it where a cat can leap on it, *don't* put it in full sun, but *do* ensure there is a perch nearby. Nest boxes in towns can be fixed to a washing-line post, an extension post to a fence, or by wire to a metal post. Security and shelter from sun and rain seem to be the main things. You will find more information in books listed on page 95. Good luck though. You could not have more delightful neighbours than a pair of nesting Blue Tits!

Mini-habitats

You can also provide habitats in your school grounds for the smaller creatures. It may take a little longer for them to be found if you only have a tarmac playground to use. Here are some ideas – perhaps you could think of others?

4:14   A piece of broken paving stone on a lawn makes a mini-habitat for the nest galleries of the Black Lawn Ant

1 Put flower pots upside-down in various places. In some put dry grass and in others damp grass. Are clay pots better habitats than plastic pots? Place flower pots upside-down on sticks, as some gardeners do amongst Dahlias. Sink a large flower pot into the ground and put a number of stones inside it.

2 Invert some jam jars on soil in different areas. Lay some on their sides – but do put a stick by them so that your friends do not tread on them.

3 Lay down some bricks and pieces of broken paving stone. Compare the inhabitants beneath those on paths and those on lawns.

4 Compare the animals found beneath damp sacking and a piece of old soft fibre-board left lying side by side.

In all these mini-habitats consider the following conditions:

(a) The amount of light available.

(b) The temperature – is it warmer or cooler than the surrounding areas?

(c) Is it damper or drier than the surrounding areas?

(d) Does the size of the mini-habitat influence the size of the animals that are found there?

(e) Does the population differ according to the seasons?

Discovery Assignment 4

1 Make a sketch map of your school grounds, marking on it the different types of habitat available. Make lists of the plants and animals you find living in each of the habitats.

2 Map the distribution of moss on your school car park. If there is more than one kind of moss present, collect samples, giving each an identification number.

3 Look for a slab of concrete that has been colonised by lichens. Make a tracing of the individual colonies. Colour in the different lichen species on your map.

4 Plot the positions of animals found on the underside of stones. Relate this to the roughness, or otherwise, of the under surface. Are more creatures found beneath stones that have been there a long time or not?

5 Consider ways in which you could make your school surrounds more attractive to birds and other animals.

5-WASTE GROUND AND REFUSE TIPS

Habitat types

In this chapter we will include as habitats any area that has been used by man in some way or another but is no longer being properly cared for. A vacant building plot, a derelict garden, a disused quarry, a reclaimed rubbish tip and land scheduled for redevelopment in a town or city are some examples. Individual soil types may differ but they will have in common some degree of soil disturbance. For a time, at least in the earlier stages of colonisation, competition between plants will not be great. Since many areas may only be 'waste' for a few months the plants that arrive must be 'opportunists'.

Colonisation and seed dispersal

To be a successful opportunist a plant must produce abundant and effectively dispersed seeds. The most striking waste-ground weeds in our industrial towns include Oxford Ragwort (*Senecio squalidus*), Coltsfoot (*Tussilago farfara*) and Rose-bay Willow Herb (*Chamaenerion angustifolium*) (Figs. 5:5, and 2:6). All these plants produce masses of wind-borne seeds. If you have a roadside gravel heap near where you live watch it regularly, recording the various plants that appear.

5:1 Soil heap colonised by Groundsel, Deadnettles, Sun Spurge, Dandelions, Fumitory and other plants. Which can you find?

5:2 The annual Sun Spurge (*Euphorbia helioscopia*) is a common plant of waste places

5:3 The showy bluish/purple flowers of the Common Mallow appear throughout the summer from June onwards

Native and alien plants

By now you should be familiar with a number of weeds with a short life cycle but which appear throughout the year (ephemeral weeds) such as Groundsel, Shepherd's Purse and Hairy Bittercress. Amongst other native plants colonising waste ground will be those shown above. The list, however, could be quite long.

Strangely enough many of the colonisers will be aliens, having been introduced to Britain from foreign lands by some means or other. Their stories are quite fascinating.

Oxford Ragwort is native to Sicily, growing very freely in the volcanic ash of Mount Etna. First noted wild in Oxford during 1794, it probably escaped from the Botanic Gardens there. It spread far and wide along walls and railway embankments until today it is found in most counties.

Hoary Pepperwort (*Cardaria draba*) (Fig. 5:4) sometimes appears in large masses. This alien is said to have been brought to Kent from Holland around 1809 in the bedding of soldiers returning from that country. It has increased steadily, some-

5:4 Hairy Pepperwort on a refuse tip

5:5 Oxford Ragwort in Worcester City (*inset* a close-up of flower)

5:6 Pineapple weed is found growing on farmland, waste places and pathways

5:7 The attractive Buddleia spreads by means of tiny winged seeds

times appearing in habitats very remote from its other locations. It may have spread in manure and fodder. One colony appeared a few years ago on the site of a gypsy encampment with horse-drawn caravans.

The Compositae family of flowering plants afford some of the most striking examples of the way alien plants have found themselves habitats on our waste land.

Where the soil has been compacted by the wheels of vehicles going to the site you may find Pineapple Weed (*Matricaria matricariodes*) (Fig. 5:6). A native of Asia, this strange-looking plant, whose flower heads smell vaguely of pineapple when crushed, is thought to have reached Britain by way of Oregon (America) during the late nineteenth century. It has been suggested that the fruits are transported by the wheels of vehicles. Collect the mud from car tyres and sow on to a flower pot of sterilised soil. Look for the seedlings of the plant.

The illustration in Fig. 5:7 shows a plant which is very successful in colonising waste ground, banks and walls. The Butterfly Bush (*Buddleia davidii*) was introduced about 1890 as a garden plant from China.

5:8 Look out for the strange plants of the Thorn-apple (*left*) with their large toothed leaves and white or purple flowers. It gets its name from the poisonous green, prickly fruits (*right*) which are about the size of a horse-chestnut case

Unusual plants

Be on the look-out for other unusual plants, for some seeds remain dormant in the soil for years and years. The Thorn-apple (*Datura stramonium*), introduced from America in the seventeenth century, was at one time quite common around London but is now rarely seen. However, it occasionally appears in large numbers all over the country when conditions are right for its germination. Such a year was 1969, when it was reported from many English counties. The author would be pleased to receive any records of the Thorn-apple (Fig. 5:8) and other plant oddities you find growing in waste places.

A look at thistles

'There's more to thistles than prickles!' Although at first you will probably think they are all alike, look closely and you will see many differences. The Spear Thistle (*Cirsium vulgare*) is one of the most widespread of our waste-ground thistles but is a biennial, whereas the Creeping Thistle (*Cirsium arvense*) is a very persistent perennial. The tall, elegant Scots Thistle (*Onopordon acanthium*) also occurs on waste ground which has been disturbed. Despite its name it is rare in Scotland!

Although they are despised by farmers and gardeners, the composite flowers of thistles are a delight to their many and varied insect visitors. The spiny black solitary larva of the Painted Lady Butterfly hides itself in the spun thistle leaves on which it

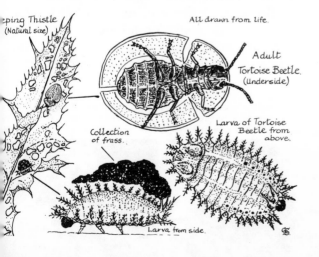

Creeping Thistle
(Natural size)

All drawn from life.

Adult
Tortoise Beetle.
(Underside)

Collection
of frass.

Larva of Tortoise
Beetle from
above.

Larva from side.

5:9   Look more closely
at the life on a Thistle

45

feeds. Skeletonised leaves on thistles may give away the where-abouts of the adult Tortoise Beetles (*Cassida rubiginosa*) and their peculiar larvae, which carry around piles of excreta on their backs, like muck-spreading fork-lifts! Quite common, too, are the roundish galls caused by a tiny fruit fly, which appear on the Creeping Thistle from July to September.

The Hogweed flower platform

The flower heads – called *umbels* – of the Hogweed (*Heracleum sphondylium*) (Fig. 5:11) seem to be a meeting ground for insects of all shapes and sizes, of which some are illustrated. Make recordings throughout the day if possible. Some insects, like Weevils eat the leaves and petals, while Hover-flies, bees and certain long-horned beetles feed upon the nectar and pollen. Some insects, such as the Social and Solitary Wasps and the furry golden Dung Flies, prey upon the nectar and pollen feeders. Here are food chains being formulated before your eyes.

Brambles as mini-nature reserves

In many parts of the country the Bramble will quickly colonise waste ground, its tangle of prickly stems, rooting at their tips, making cover for birds. Its white flowers, like those of the Hogweed, provide food for many insects that meet to mate.

The leaves provide a mini-habitat for the larva of the Leaf Mining Micro-moth (*Nepticula aurella*) (Fig. 5:10.)

When the fruits form in autumn there is again an abundance of visitors, from wasps, flies, butterflies, to birds and Badgers. From your observations you could make up a food web similar to that in Fig. 5:12 which has been partly worked out from the collection of photographs made by the author whilst studying the Bramble bush in autumn.

5:10 Gallery of the
Bramble Leaf Miner

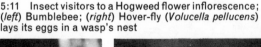
5:11   Insect visitors to a Hogweed flower inflorescence;
(*left*) Bumblebee; (*right*) Hover-fly (*Volucella pellucens*)
lays its eggs in a wasp's nest

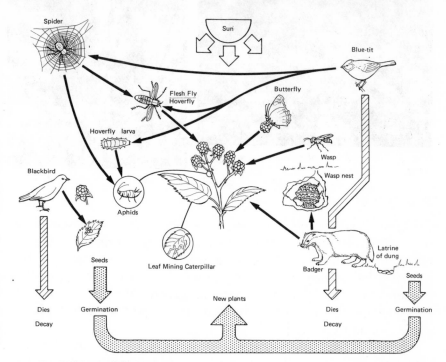

5:12 Food Web of a Bramble bush

Birds and bird pellets

Quite often you may come across collections of seeds, fruit skins and purplish froth on Bramble leaves. These have probably been ejected from the bill of a Blackbird, a phenomenon known as 'pip-spitting', which is a method of seed dispersal. Many birds, however, regurgitate easily recognised pellets of material which has not been digested (Fig. 5:13). Be on the look-out for such examples, for when pulled apart they give an indication of the food on which the bird has been feeding. Gull pellets are likely to be found on old rubbish tips.

5:13 Bird pellets to look for

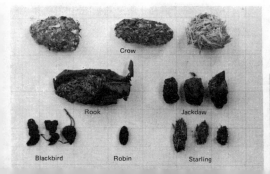
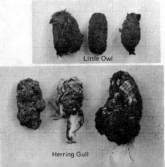

5:14 Larvae of Peacock Butterfly on stinging nettle

5:15 Galls of the Nettle Gnat

A patch of nettles

Stinging Nettles (*Urtica dioica*) frequently become established on waste ground and rubbish tips, especially in areas where there is an accumulation of decaying waste material. Despite their stinging hairs they provide a habitat for many creatures. The presence of the brilliant metallic green Weevils (*Phyllobius pomaceus*) is given away by the vast numbers of small holes in the leaves made by their jaws.

Some of the aphids may appear as a brown shell with a tiny hole (Fig. 1:4). These will have been parasitised by a tiny Chalcid Wasp. Ants may be found 'milking' their herd of aphids. Look also for the butterfly larvae in Fig. 5:14 and for leaves rolled and secured with silken threads. These may contain larvae of either the Mother of Pearl Moth, or of the Small Magpie Moth. Some Nettle leaves and stems may have galls on them caused by the Nettle Gnat (*Dasyneura urticae*) (Fig. 5:15).

Earwigs as mothers

If you are looking below stones in the winter months you may disturb a female Earwig with a clutch of whitish spherical eggs.

5:16 A female Earwig tends its eggs and newly hatched young

5:17 The Minotaur Dung Beetle searches for rabbit droppings

These she will guard, licking them from time to time, until they hatch into soft, pale-coloured miniature Earwigs (Fig. 5:16). After a few moults the young Earwigs eventually disperse. This apparent care of eggs and young is very rare in insects. You may, however, also come across a species of Shield Bug which has a similar habit.

Nature's refuse collectors

In every habitat there are members of Nature's own sanitation department, feeding upon dead plant and animal material.

The bluish-black bulky Dor Beetles (*Geotrupes*) are often seen with their undersides infested with a species of mite whilst feeding on cow dung. The Minotaur (Fig. 5:17) feeds upon rabbit dung. Both dig deep tunnels below the ground, provisioning them with dung, on which the eggs are laid.

However, it is the Burying (or Sexton) Beetles (*Necrophorus species*) which have achieved such fame in Nature's sanitation department. These are likely to be found burying a dead bird, a Shrew and yes, even tackling a dead Badger! The eggs are then laid in the carcass, the beetles remaining to feed the grubs upon regurgitated food! After a short time the grubs feed upon the carcass themselves.

Nothing is wasted. Even the hair is eaten by specialised beetles, moths and their larvae. The driest of bird pellets will disintegrate after the Museum Beetle (*Anthrenus museorum*) has tackled it.

Discovery Assignment 5

1 Collect the seed heads of Coltsfoot and Oxford Ragwort. Count the seeds in a head. Calculate the seeds produced per plant.
2 List the plants colonising a piece of waste ground. By using one of the described sampling techniques discover which plants are the commonest. How might the seeds have arrived?
3 Collect the galls on Creeping Thistle in September, keeping them in a container until the adult flies emerge during the following spring. Are there any parasites?
4 Rear up in containers the larvae of the Mother of Pearl and the Small Magpie Moths found on Nettle (see page 91).
5 Find out more about the life cycles of Dung Beetles, and also the various flies that feed upon decaying material.

6-WALLS-OLD AND NEW

Walls of Victorian gardens, city boundary walls, walls between houses, walls instead of hedges in the country, even the walls of your house and school; all provide plants and animals with a place to live. They may not be noticeable at first perhaps but they are there all the same.

Wall types

For countless centuries man has been building walls. The Great Wall of China and Hadrian's Wall across the Pennines are but two examples. Can you think of any others? All over this country there are to be found examples, but quite often the more interesting ones will be those of old castles, around walled-in gardens or in the older parts of our cities and towns. Walls are used to mark boundaries and to prevent people and animals from straying over adjacent land. Dry-stone walling has occurred since the Iron Age. In places it was easier to pile up stones collected from fields than to dig and pile up earth as early field divisions. Look at the picture of the dry-stone walls separating the field divisions of the 'Viel' (pronounced 'Vile') near Rhossilli in the Gower Peninsula and compare it with that of the brick wall by the River Severn in Worcester City. Have they anything in common?

Building materials for the wall

Quite often the stone used in walls is fairly local and may, like that of the Gower walls, have been dug almost on the spot, yet some stone indicates it may have come from elsewhere.

6:1 Brick and sandstone city boundary wall, Worcester

6:2 Dry-stone walls of carboniferous limestone at Rhossilli

Walls may be of limestone, granite, slate, sandstone, boulder and flint, and of course brick. The weathered surface of limestone may show evidence of fossils, such as the 'sea-lilies' or crinoids Perhaps the stone has a local name such as the Lancashire 'Haslingden Flag' once used for flagstone pavements of the industrial areas of Rochdale and Bolton.

Even an ancient brick wall may have its local associations. Old clay bricks are often irregular, having been roughly fashioned by hand. The wall shown in Fig. 6:1 had a number of bricks that had been over fired. The sizes of the bricks too were strange compared with those now processed in electrically operated kilns. The mortar used for bonding the bricks together differs from that used today.

Lichens and air pollution

Of all the living things that may be found on a wall, lichens are the strangest. The crusty looking patches of grey, yellow or orange are lichens, a combination of two plants in one. An alga and a fungus have developed a remarkable way of living together so closely that they appear as one. Each plant exhibits a sort of mutual help called symbiosis. Because this mutual help is so successful lichens are to be found everywhere in the world except where the air is polluted by industrial smoke. Lichens with few exceptions are very susceptible to atmospheric pollution, particularly to sulphur dioxide. Much of their food is obtained dissolved in rain water, which falls upon the surface of the lichen. Thus the poisonous substances also in the rain water accumulate in the algae cells, resulting more often than not in the death of the lichen (see *Pollution and Life*).

Two types of lichens are to be found on most walls – 'crustose' and 'foliose' (Fig. 6:3). By looking at lichens beneath a hand lens

6:3 Part of Velsh wall showing (*a*) crustose, (*b*) foliose, and (*c*) fruticose

you will be amazed at the detail of structure, opening to you a whole new world. You may even see small creatures, such as *Collembola*, grazing upon the spores liberated by the lichens.

Algae, Liverworts and mosses

The wall habitat may be like a desert during the hot summer months so that much of the plant life will have to grow during the wetter winter and early spring months.

Where a wall is shaded by the branches of a tree it will quite often be covered by a pale green powdery film – an alga called *Pleurococcus*. This is similar to that which is found on the moister side of tree trunks.

Liverworts of course will only be found in areas of a wall which is continually moist. Liverworts were so called because certain types were in the past thought to resemble the lobes of a liver. A plant bearing a resemblance to a human organ was thought to contain substances helpful in curing diseases of that organ. The two Liverworts most likely to be found in towns are *Lunularia cruciata* and *Marchantia polymorpha*.

Mosses, however, are often the early pioneers into the hostile environment of the wall. Because they hold water, and keep the wall surface beneath them moist, they assist in the breaking down of the brick or mortar surface. Look for the small neat cushions of the Wall Screw Moss (*Tortula muralis*) on the tops of walls, and also the much more domed tufts of the Grey Cushion Moss (*Grimmia pulvinata*), which will also be found on stone window-sills and roofs (Figs. 4:2, 6:4 and 6:5).

6:4 Colonisation by the Wall Screw Moss (*Tortula muralis*)

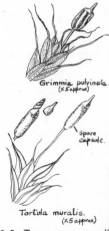

Grimmia pulvinata.
(×5 approx)

Spore capsule.

Tortula muralis.
(×5 approx)

6:5 Two common wall mosses in close-up

6:7 Hartstongue Fern and the smaller Wall-Rue Spleenwort Fern

6:6 Maidenhair Spleenwort Fern

Ferns

Before walls existed most of the ferns that we find on walls grew in crevices of cliffs where particles of soil could be trapped. It is amazing how ferns grow with rootstocks penetrating the cracks and crevices of brick walls. One wall at Worcester contained four species (Figs. 6:6–6:8), whilst another, the Rusty-back Fern (*Ceterach officinarum*) was discovered on a nearby railway bridge (Fig. 6:9). There are a number of interesting local names for this fern such as Brown-back, Scale Fern, Stone Fern and Stone-breaker. It can survive in the driest of walls, growing during the winter but with its leathery leaves becoming dry and brittle during the summer months. Possibly the most widespread wall fern is the little Wall-Rue (*Asplenium ruta-muraria*) (Fig. 6:8). Ferns do not form seeds but reproduce by liberating masses of tiny dust-like spores.

6:8 Black Spleenwort Fern

6:9 Rusty Back Fern

6:11 The heavy seeds of this Yellow Corydalis are probably dispersed by ants

6:10 Ivy-leaved Toadflax grows amidst the tiny leaved Mother-of-Thousands

Flowering plants of walls

Some plants you find growing on walls will have escaped from nearby gardens, such as the Antirrhinum and Rock Arabis, whilst the Buddleia mentioned on page 44 is often found in large numbers. Some, too, may be common weeds such as Groundsel, Dock, Dandelion and Plantain, but others may be species only to be found on the wall habitat. A number of plants have included in their English or Latin name some reference to their habitat – 'wall' or 'muralis'. A few examples are Wallflower, Pellitory-of-the-Wall, Wall Pennywort and Wall Speedwell. Look through a wild-flower book for more.

One of the most widespread wall plants is the dainty Ivy-leaved Toadflax (*Cymbalaria muralis*) (Fig. 6:10). This successful alien plant effectively clothes large areas of wall with its delicate long trailing stems, small ivy-shaped leaves and attractive purple and yellow snapdragonlike flowers. Once its flowers are pollinated they bend towards the wall so that when the capsule discharges its tiny seeds they are sent flying into any convenient crack.

Pellitory-of-the-Wall, a member of the Nettle family, is a true native of Britain. The insignificant clusters of greenish-brown flowers are unusual (Fig. 6:12). When an insect touches the flower, pollen is discharged in a 'puff', dusting both insect and adjacent flowers. Because of this device is sometimes known as the 'Artillery Plant'.

6:12 Pellitory-of-the-Wall on a church wall

6:13 Wall Pennywort, sometimes called Navelwort, grows in damp, shady parts of walls

Place a thermometer against a south-facing wall in summer and you will be surprised how hot it is. Thus some plants have specially thickened fleshy leaves for conserving moisture. Examples are Wall Pennywort and Wall Pepper (*Sedum acre*) (Figs. 6:13 and 6:14).

Mother-of-Thousands (*Soleirolia soleirolii*), also known as *Mind-your-own-business*, is a plant which has escaped from cold greenhouses to the wall habitat where it forms a mat of small rounded leaves (Fig. 6:10). Strangely enough it is related to the Pellitory-of-the-Wall.

6:14 The succulent-leaved Wall Pepper has yellow flowers. The larger-leaved English Stonecrop has white flowers

6:15 The underside of the divided leaves of Mugwort (*Artemisia vulgaris*) appear silvery

Wall plants in history

You may have noticed that a number of wall-plant names end with '*wort*' in the English (Spleenwort, Pennywort, etc.). In Anglo-Saxon times a favoured medicinal plant was indicated by 'wort'. Later on, around the sixteenth century, when various herb plants were being used by the monks or friars for inclusion in their 'official' prescriptions, the plants were given the name '*officinalis*'. Culpeper's *Complete Herbal* gives a number of examples, such as *Parietaria officinalis* (Pellitory-of-the-Wall), a treasured drug plant used to 'Comfort the body and provide a remedy for sore throats and also bruises'! Mugwort and Wormwood (Fig. 6:15) are just as likely to be found on waste ground as on walls. For hundreds of years they have been treasured as important plants. 'They that travel, if they carry Mugwort, will never tire.' In the Middle Ages Wormwood was hung up over the door and used for getting rid of intestinal worms!

> What savour is better, if physicke be true
> For places infected than Wormwood and Rue.

Listing wall plants

Make lists of plants seen on walls of different materials, of different heights, those that are shaded and unshaded, as well as those damp and very dry. Compare lists made in towns with those in the country. The following might help:

Wall details..................		Location............	Date............
Position on wall	*Plant species*	*Annual/ Perennial*	*Seed dispersal method*
Top of wall			
Ledge (if present)			
Mortar			
Brickwork			
Junction of wall and base			

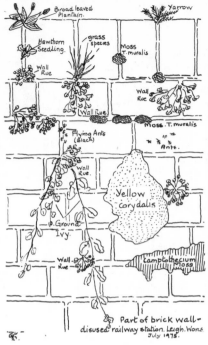

6:16 Belt transect of an old wall

6:17 Old castle walls are habitats for many plants, including the yellow Wall-flower

Mapping the distribution of wall plants

The illustration shows a sketch map of part of a wall. Quite often a narrow belt does not show all you require, but this could be supplemented by a number of individual squares made at different parts of the wall. It is possible to make a 'rubbing' of a brick wall on a length of kitchen paper. This can be used for marking on the sites of individual plants. On these maps you can superimpose the positions of any animals found.

Animal life

Walls generally are very hostile environments for animals. Most of those present will be found beneath the protection of the flowering plants, although tussocks of moss often conceal animals like *Collembola*. A search should be made, using a hand lens or binocular microscope.

6:18 *above left* Osmia Mason Bee on wall of Grimley Primary School, Worcestershire

6:19 *above right* Untidy web of Ciniflo Spider amongst fronds of Hartstongue Fern

6:20 Zebra Hunting Spider on brick wall

A sunny wall is often a resting place for flies. Where there are flies there are usually spiders! The little black-and-white-striped Zebra Hunting Spider (*Salticus scenicus*) (Fig. 6:20) is likely to be seen, and of course the bluish-grey untidy webs of the Sieve and Comb Spider (*Ciniflo sp.*) (Fig. 6:19).

Old walls particularly may be colonised by a number of the fascinating wall-living bees and wasps. Most common of these solitary bees, and even to be found living in new walls, is the one shown in Fig. 6:18. *Osmia rufa* is the commonest species of the 'Mason-bee' group. They are called this because they form their cells of tiny particles of earth, sand and minute pebbles, all glued together by their saliva. Each cell is stocked with provisions of honey and pollen before an egg is laid.

However, it is not always an *Osmia* bee that emerges early the following summer. Sometimes a brilliant-coloured Ruby Wasp (*Chrysis sp.*) appears instead, telling us that during the previous year a female Ruby Wasp must have slipped into the crack unnoticed and laid an egg inside the bee cell before it was sealed up. The wasp grub would have devoured the bee grub and the provisions collected by the unsuspecting *Osmia* bee. Whereas bees are vegetarians, wasps provide only animal food for their offspring. The cells of the small black and yellow Mason Wasp (*Ancistoucerus parietum*) are stocked with small caterpillars. These'have been paralysed by the wasp's sting in order that they remain alive and fresh until the wasp grub is ready to eat them! The Mason Wasp, as well as using holes in walls, quite often uses a door lock!

Ivy as a wall micro-habitat

Often the main plant coloniser of a wall will be ivy. Ivy is both a food plant for various creatures and also a micro-habitat for the numerous animals which find shelter beneath the thick cover of evergreen leaves. Search carefully, recording the animals you find. You may discover two or three species of Woodlouse, a number of different Snail species, *Collembola* and also primitive three-pronged Bristle-tails (Silverfish) mentioned on page 12. Most of the creatures you find will be those requiring a high degree of moisture and shade. Many only move around during the hours of darkness.

Ivy blossom, which appears in late autumn, is always eagerly visited by moths, Bluebottle Flies, Hover-flies and the Social Wasps, for the rich supply of nectar. At this late time of the insect year many are lethargic and drowsy. Observe how long each spends upon the flower head, how they feed and groom themselves.

Discovery Assignment 6

1 Find out where the older walls in your locality occur. By going to a reference library look at old street maps and street directories to discover what the area was like in the past.
2 Make a survey of the plant and animal life on the walls surrounding a churchyard. By comparing the different lichen types which are growing on grave headstones with those on the wall, can you give an approximate age for the wall?
3 After getting permission, attach a length of sticky greaseband paper to a wall. This will enable you to find out how much soil and how many different plant propagules (fragments of plants) arrive in a given time.
4 Find out more about the ferns of your neighbourhood and how to grow them. What other kinds of ferns are sold in shops?
5 Make a 'bottle garden' out of a sweet jar.

7-HEDGEROWS AND VERGES

The hedgerow

Hedges are a unique part of the countryside as well as being a haven for wildlife. Have you ever wondered how the hedge became such an integral part of the English country scene?

Hedges in history

Looking at Fig. 7:1 it is perhaps difficult to realise that there were very few hedgerow boundaries before the Parliamentary Enclosure Acts of the late eighteenth century. Whereas the pre-enclosure hedge is usually winding, the post-enclosure hedges are straight in line and are more likely to consist only of the quick-growing Hawthorn (sometimes known as Quickthorn).

Maps and hedges

The landscape is constantly changing. Looking at a map of your own area is like taking a step backwards in time, for over the last twenty or so years there has been widespread removal of hedges,

7:1 Aerial view of farmland showing hedgerows linking small particles of woodland

7:2 and 7:3 Sketch maps showing how the same piece of land has changed during a period of time. The loop of the river Teme is about the only feature unchanged

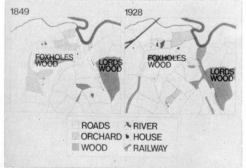

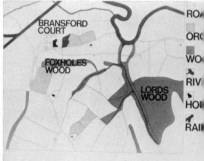

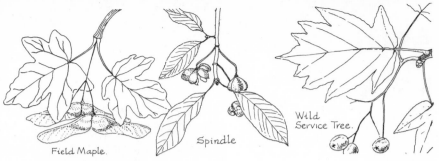

Field Maple.

Spindle

Wild Service Tree.

7:4   How old is your hedge?

particularly in the arable areas of eastern England. By comparing the field boundaries on old maps with those of the present time you might be able to show something like that in Fig. 7:3.

Dating a hedge

It is possible to date a hedge roughly by counting the different hedge shrubs and trees growing in an average 30 m stretch. By using hedges that could be positively dated by documentary evidence, it has been concluded that a hedge is colonised by one new species approximately, each century. It has also been found that certain shrubs seem to have occurred in a hedge when it was a given age. Field Maple (*Acer campestris*) occurs in hedges which are 400 years old, whilst Spindle (*Euonymus europaeus*) occurs in hedges some 600 years old (Fig. 7:4). The Wild Service Tree, also in Fig. 7:4 is in itself an indicator of very ancient woodlands, so it is reasonable to suppose that a hedge containing it will be very old too.

Hedges as corridors for wildlife

A large part of the animal life of our country must have originally belonged to woodlands. As the woodlands diminished and hedgerows increased, so many woodland creatures moved into the hedgerows. Recent investigations show that of the forty or so species of birds to be found on ordinary farmland, most are to be found in association with hedges. Some species, like the Whitethroat and Yellow Hammer (Fig. 7:5) are almost always found nesting in hedgerows. As you can see from the aerial photograph (Fig. 7:1), hedges connect the remaining areas of small copses and groups of trees like a series of corridors. Were it not for these,

61

7:5 Yellowhammer feeding young

7:6 Newly layered hedge in early summer

animals such as the shy Bank Vole (*Clethrionomys glareolus*) would only be found in woodlands.

Hedge management

The art of hedge layering or 'pleaching' (Fig. 7:6) was practised to rejuvenate old tall and straggly hedges, making them stockproof again. Notice the intertwined rods, often of hazel, along the top. This, called 'hethering', prevents cattle from pushing the newly laid hedge out of position. Hedge laying is time consuming and thus expensive. How long will it be with us?

In the past the hedger and ditcher used a very sharp, long-handled 'slasher' for the hedge, and a spade for the ditches. Nowadays it is done by machine, for a machine can lop a hedge at a walking pace, whereas the hedger could probably only lop 100 m of hedge in a day. However, the machine does not clear the old dead wood and other material from the hedge base, to allow in light and air. Will this, in time, affect the flora of the hedgerow?

Galls and their occupants

When a hedge is cut back hard, the subsequent growth is often quite spectacular. It is on the new growths of oak that one sees the strange woody marble galls, often in large numbers, that are caused by a tiny wasp-like insect (*Andricus kollari*). The Marble Gall Wasp lays eggs in the terminal or lateral leaf buds of oak. Each larva, for there is only one per gall, causes the oak tissues

7:7 Female Gall Wasp (*Andricus kollari*) just emerged from oak marble gall (much enlarged)

to grow around it, thus providing itself with nutritious food. After a short period as a pupa the adult insect emerges, by biting its way out of the gall, usually in September (Fig. 7:7). Sometimes the rightful occupant is parasitised by Chalcid Wasps, and these in turn are parasitised by other Chalcids which emerge during late March.

Other galls likely to be found on hedgerow trees and shrubs are shown in the various illustrations (Figs. 7:8 to 7:11).

7:8 Oak apple galls

7:9 *right* Bedeguar gall on wild rose

7:10 Spangle galls on oak leaf

7:11 *right* Bean galls of Sawfly on willow

Plants of the hedge

Many of the flowering plants which grow in association with the hedge are also likely to be encountered in the lighter parts of deciduous woodlands. Examples of such woodland plants are Dog's Mercury, Bluebell, Primrose, Sweet Violet, Wild Arum, Wood Goldilocks and Wood Spurge (Fig. 7:12). All these are spring-flowering perennial plants.

Of the hedgerow climbing plants, all are perennials with the exception of Goosegrass (*Galium aparine*) which is an annual. Black Bryony, Bittersweet and White Bryony (Fig. 7:13) are generally more abundant in managed (sometime layered) hedges. The presence of Honeysuckle is more marked in hedges which have not been recently layered. Wild Rose (both *Rosa canina* and *R. arvensis*) is included in the shrub list for dating purposes.

Some flowering plants seem, by virtue of their names, to have found a permanent niche in the hedgerow habitat. Hedge Garlic (*Sisymbrium officinale*), Hedge Woundwort (*Stachys sylvatica*) and Hedge Bedstraw (*Galium album*) are but three examples.

Hawthorn and its animal life

There are probably more hedges of the common Hawthorn (*Crataegus monogyna*) in this country than of any other shrub. It also seems to have more local names than many of our native plants. Do you know it as: May tree, bread-and-cheese tree, whitethorn, peggles, hipperty-haw tree, tramp's supper, hawzey bush, moon flower, or as something else? This is one of the reasons why it is necessary to have one Latin name!

Hawthorn is a food plant for the larvae of over eighty species of moth, of which about one-third do not have any alternative food apart from the other hedgerow shrubs of Blackthorn and Rose. The moth larvae make use of a number of habitats in the Hawthorn. Some, such as the Gold Tail Moth (Fig. 7:15), the Vapourer and the Lackey, feed upon the leaves in full view; some, such as the various Winter Moth species, spin leaves together; yet others, like the various micro-moth larvae of the genus *Coleophora*, mine inside the leaves. There is one interesting species of these that moves around the leaf, its body encased by particles of leaf. When a Hawthorn hedge is not too closely clipped it will produce an abundance of white, sometimes

7:12　Wood Spurge

7:13　White Bryony on town hedge of *Lonicera nitida*

7:15　Larva of the Gold Tail Moth

7:14　Reddish/purple flowers of Hedge Woundwort and leaves of the climbing Black Bryony

7:16　Larvae of the Lackey Moth on their communal web

7:17　Honeybee on Hawthorn flowers, seeking nectar and pollen

7:18 The flattened Hawthorn Shield Bug has sharp beak-like mouth-parts

pink-tinged, flowers. Throughout the days of early May the flowers attract large numbers of insects, including Bees, Bumble Bees, Hover-flies, Weevils, Click Beetles and Longhorn Beetles.

Feeding upon the leaves and rich red berries with its sharp needle-like proboscis will be the common Hawthorn Shield Bug (*Acanthosoma haemorrhoidale*) (Fig. 7:18). The haws are also attractive to birds, especially members of the Thrush family. They are also an important food for the long-tailed Wood Mouse (*Apodemus sylvatica*) which often feeds inside an old Thrush's or Blackbird's nest, leaving the chewed remains of fruit and seed.

The roadside verge

Roadside verges and their hedgerows, which criss-cross the British countryside provide a refuge for much of the plant and animal life of the district. It has been said that their total area, some 440,000 acres (180,000 hectares), is twice that of the whole of the nature reserves in this country. A leading naturalist has suggested that about seventy per cent of the plants listed in a British wild-flower book can be found within a few metres of the roadside. Records collected by the national Biological Records Centre show that of the 300 rarest plant species to be found in Britain, the main habitat for twenty-seven of these is on roadsides. More exciting still, seven of these have been discovered growing on roadside verges alone. This means that we must show great care when working on roadside verges.

7:19 An attractive wide verge alongside a busy Suffolk road

Verges as grassland habitats

The roadside verge often presents various forms of unfertilised permanent grassland, unploughed for perhaps hundreds of years. Some may go back even to Roman days, but more of course may have been formed by the Turnpike Trust in the nineteenth century. Even so, because they must cross all the geological formations and varied soils of a county, they must provide a comprehensive grassland flora of the county. Some verges will be damp, others well drained; some will be sunny and others shaded; some will be exposed, others sheltered, and some will be flat or sloping. Such a variety in a county makes them well worthy of study. Of course not all verges will be attractive – some may be so narrow or so disturbed that few plants or animals are able to exist on them.

The plants on the average verge will have to be pretty tough to live alongside a busy road. They have to withstand exhaust fumes from motor vehicles, salt spray during the winter months, and other disturbances. Amongst the most successful plants are those that can reproduce themselves very freely. Many of these are often loosely called 'dandelions' by the average person, for under that heading they lump together all the yellow composite flowers like true Dandelions, Coltsfoot, Hawkweeds and Hawkbits. Many of the latter are very difficult to distinguish. You don't have to go far to find these either. Quite often a wide verge in a new housing estate will provide vast numbers.

Roadside butterflies and verge management

A number of the commoner roadside butterflies have larvae which feed upon various rough grasses. The Meadow Brown,

7:20 Throughout the day Dandelion flowers attract bees, but close up at night

7:21 *top, right* Ringlet Butterfly

7:22 Meadow Brown Butterfly

Hedge Brown (or Gatekeeper) and Ringlet are probably the most widespread (Figs. 7:21 to 7:23). Where the soil is dry and sandy, the Small Heath and Large and Common Skippers may be seen amongst the coarser grasses. The larvae of the delightful Small Copper Butterfly feed upon Sorrel, and those of the Common Blue upon Bird's-foot Trefoil and various Clovers. Many of the grass-living butterflies over-winter either as hibernating larvae or as a pupa attached to a grass stem. If the whole of the verge was mown there would be little chance of these surviving. However, when we remember that up to fifty years ago many of the wider verges were used by the local cottagers to graze their animals we can perhaps understand why it is necessary for our verges to be managed in some way. Of course the main reasons of the various councils' management of roadside verges are not always the conservation of the flora and fauna, but concerned with the safe use of the roads for traffic. In some counties stretches of particularly interesting verges are marked by special posts such as you can see in Fig. 7:29.

Adopting a verge for seasonal observations

Try to find out how long the verge has been there. Remember that sixty years ago or less many roads, even quite important ones, had a loose and dusty surface flanked on both sides with a

7:23 Hedge Brown (or Gatekeeper) Butterfly

7:24 Flail-type rotary verge cutter mounted on a tractor

7:25 The star-shaped composite flower of the Milk Thistle

7:26 How common is Agrimony in the verges of your district?

7:27 The Meadow Cranesbill is now mainly found on roadside verges

7:28 Verge containing Cowslips and the Green-winged Orchid (*Orchis morio*)

strip of ground. The general adoption of tarmac, concrete and other forms of 'metalling' the surface may have had some effect upon the roadside vegetation. Perhaps seeds, or parts of plants, were accidentally introduced from other areas?

Up to ten or fifteen years ago most roadside verges were cared for by men called 'lengthsmen', whose job it was to look after a length of road and the adjacent verges, ditches and hedges. As these workers retired so their duties were taken over by men with machines (Fig. 7:24). Find out how your verges are managed. Different treatments will encourage different plants and animals.

The making of a number of plant strip-maps from one side of the verge to the other should show the effects of both management and road traffic. By recording the plant and animal life throughout the season you may be able to suggest conservation objectives for your adopted verge.

Adopting a newly seeded verge

Perhaps a road near you has recently been straightened out to get rid of a dangerous bend, and a new verge laid down. This would make an ideal study area to observe and record the various stages in its natural colonisation. Find out what kind of seed mixture was used initially, and when it was done. One such newly seeded verge just outside Worcester was carefully observed and at least twenty-five plants appeared which had not been recorded before in the immediate neighbourhood. Of these new species, five were 'aliens', including the Milk Thistle (*Silybum marianum*) in Fig. 7:25 and three species were not even on the British list of flora! Of course some seeds could have been brought in on the wheels of the contractors' vehicles and others could have been dormant in the soil, but it does illustrate how much there is to find out about even the newest of our roadside verges.

Surveys of specific verges

Instead of adopting a verge why not adopt a specific plant of roadside verges? Find out as much as possible about the plant, its life cycle, and on which verges it is to be found growing. Agrimony (*Agrimonia eupatoria*), Meadow Cranesbill (*Geranium pratense*), Cowslip (*Primula veris*) and Pyramidal Orchid (*Anacamptis pyramidalis*) are some examples (Figs. 7:26 to 7:28).

7:29 A roadside Nature Reserve in Worcestershire. The marker post indicates that this stretch of verge should be left uncut

Motorway verges

The natural colonisation of motorway verges, embankments and cuttings is very interesting, although we can only look from the window of a car. It is quite likely that these new habitats will provide sanctuaries for many plants and animals too that would not otherwise be protected. More plant cover produces more small mammals, such as Voles and Shrews. There seems to be an increasing number of Kestrels taking advantage of this new food supply along many hundreds of miles of motorway.

Discovery Assignment 7

1 Make a survey of the hedges in your own area to find their age. By comparing the hedges of today with those shown on an old map, work out how many miles of hedgerow have been removed.
2 Collect some galls and try rearing out the adult insects.
3 Adopt a length of hedge and list the plants which are found:
(*a*) On both sides of the hedge.
(*b*) On only one side of the hedge.
(*c*) On only the opposite side of the hedge.
Mark which way the hedge faces.
4 Make a profile diagram and strip-map of part of a hedge (see Teacher's Guide Book 4, page 81 for details).
5 Find out the percentage germination of seeds collected from roadside verge plants. With permission from the County Council and County Naturalists Trust you might be able to plant out the seedlings on to your adopted verge. Details of the national scheme may be obtained from the Botanical Society of the British Isles (address see page 95).
6 Create a wild-flower nursery at home or at school.

8-OPEN SPACES, COMMONS AND GREENS

The history of our commonland

Long before the days when fields near the village were enclosed by hedges, the villagers were able to use the poorer areas of ground 'in common' with each other. There they were able to graze so many sheep, cattle, pigs, horses and geese, as well as having the right to collect firewood, fencing wood, bedding for animals and so on. On some commonland these rights still remain, although in many cases they are not used.

Generally however, the commons, which may be heathland, scrub/woodland, rough pasture, fenland or a mixture of all, have become places for public recreation.

The history of each village green may be different, and may also be subject to local by-laws.

Mostly about grass

You may think there is nothing very special about the common or green. They are places for you to play on, for horses to gallop over, and often for passing motorists to find a place to park and picnic. Because the vegetation has been grazed on and off for centuries the area will mostly be grass and Bracken, with perhaps a few stunted shrubs and Birch trees.

Yet these grassy areas are important; indeed the grass family is the largest and most important family of flowering plants in Britain. We should get to know them better. Different kinds of grass are to be found growing in different habitats. Those growing on heaths will be different from those growing in woodlands.

8:1 Sheep grazing on Bringsty Common, Worcestershire

8:2 The result of cars parking on Cavendish Green, Suffolk

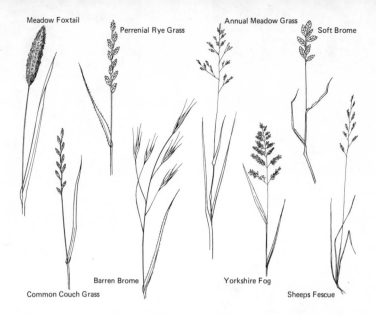

Meadow Foxtail

Perrenial Rye Grass

Annual Meadow Grass

Soft Brome

Barren Brome

Yorkshire Fog

Common Couch Grass

Sheeps Fescue

8:3   How well do you know your grasses?

Many of the grasses of commons will be those associated with poor, infertile land. It is on the roadside verges that we are likely to find some of the agricultural grasses. Nevertheless, all of them, when in flower are extremely attractive in their own right.

When you look closely at any area of grassland you will surely find other plants growing amongst the grasses. What better place to study these initially but around the school or at home?

Associated weeds

The village green is probably more similar to a lawn or playing field. Yet even here you will be surprised to find that the associated weeds differ in say the ornamental lawn, and the playing field. Herbs such as Yarrow, Ribwort Plantain, Dandelion and the various Clovers, by tapping minerals deep in the soil, provide a valuable addition to the food available in the grass.

Grass on the whole is fairly shallow rooting, whereas plants with long tap roots may penetrate the soil surprisingly deep. Examples of such roots were dug up complete: a dock root was 94·5 cm long whereas a dandelion root was a mere 53·5 cm.

An analysis of the broadleaved weeds of different grass areas may be simply carried out by throwing a square of wire (say 10 cm²) at random, and recording each time a species of weed

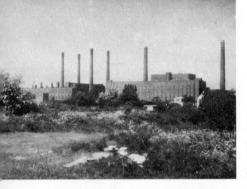

8:5 The biennial Foxglove also colonises bared ground

8:4 *above left* The soft and spineless gorse seedlings often appear after a fire

8:6 Stourport Common, Worcestershire is a Site of Special Scientific Interest (S.S.S.I.) but even so, look at the rubbish left by visitors!

plant occurs. One old lawn which at one time was part of a Victorian walled-in garden contained sixteen different weed species as well as a number of mosses!

If you are working on a common, a quick walk round before you start might enable you to divide it up into a number of distinct areas. These could be marked on a large-scale map and later analysed separately. You may have areas of mixed grassland with Bracken, containing Sorrel, Rose-bay Willow Herb, Lesser Stitchwort and Gorse. There may be Bramble areas which may contain seedling trees of Hawthorn, Birch and possibly Elder. The more open grassland may contain as grass species those illustrated in Fig. 8:3. These distinct areas illustrate the changing character of grassland, showing how, if left to itself it would pass back to some form of woodland as it was before man cleared it in those far-gone manorial days. When this happens in a habitat we call it 'succession'.

Effects of man's activities

Certainly there will be areas showing the activities of man. There may be places where horses have dunged and trodden the ground quite heavily. There may have been an accidental fire, perhaps allowing seedling Gorse to regenerate (Fig. 8:4), or there may be a place where some rubbish has been purposely burned. Make lists of the plants of these areas – for each has become a habitat quite different from the main part of the common. One of the early colonisers of such bared areas is the biennial Foxglove (*Digitalis purpurea*) (Fig. 8:5).

Look particularly for spots where cars have parked and worn away the grass on village greens (Fig. 8:2) and record the change in the flora. Look around those areas for bottles and tin cans thrown away by thoughtless people. They may think they have hidden them from view but these glass and metal containers can become coffins for the small mammals of the common. The narrow openings are often irresistible to the ever-inquisitive Voles, Shrews and Mice, for they squeeze through the openings and become trapped inside, to die of starvation or fright (Fig. 8:9).

Life amongst the grass roots

For many creatures the grass provides both food and protection, but for others it is a grassy jungle through which they must struggle, seeking out and stalking their prey. Below the soil though, various larvae are feeding on the grass roots. The thin, shining brown larvae of Click Beetles (*Agriotes obscurus*), known as Wireworms, may often be found in large numbers (Fig. 8:7). The fat, bloated, legless bodies of the Leatherjackets are to be found in moister areas (Fig. 8:8), unseen until September when, as strange-looking horned pupa, they struggle to reach the soil surface. There the adult insects emerge as the gangling long-legged Daddy-long-legs or Crane-flies (*Tipula paludosa*) (Fig. 8:8).

8:7 Grass-root feeders – Wireworms

8:8 The legless Leatherjacket larvae pupate amongst grass roots (*left*) from which emerge the adult Crane-flies (*right*)

8:9 Small mammals are trapped inside bottles thrown away by man – this Common Shrew was later set free

8:10 A young Short-horned Grasshopper. The wings are not completely developed

Using a sweep net

To obtain some idea of the myriads of tiny insects inhabiting the grassy jungles it is fun to use a sweep net for obtaining samples from different places. A sweep net is really a tough calico or nylon bag around a wire frame, attached to a short handle. This enables us to 'sweep' the vegetation from side to side as we progress forward. The catch can be either put into small boxes or sucked up into a 'pooter' (Fig. 8:13).

The grass feeders and dwellers

As well as the numerous grass-sucking bugs and aphids, a number of other interesting creatures may be found. Short-horned Grasshoppers will be found in sunny places. Since they are insects of warm countries, the warm sandy soil of many commons provides suitable conditions for their life cycle.

A number of the butterflies likely to be found basking on some flower platform were mentioned on page 68. The Grayling Butterfly is a characteristic species of dry stony commons with rough grasses, where it seems to disappear from sight after alighting upon the uneven ground (Fig. 8:14). The larva of this butterfly too is a grass feeder.

One of our commonest and most showy day flying moths is the Six-spot Burnet (Fig. 8:11). The larvae, although they are feeders

76

8:12 Male and female Bloody-nosed Beetles mate amidst the golden flowers of the gorse

8:11 A Six-spot Burnett Moth dries its wings, after having just emerged from its papery cocoon

of Clover and Trefoil, frequently produce their thin papery cocoons attached to grass stalks. Walking through the grass in early July we may disturb hundreds of the strange Grass Rivulet Moths.

Earlier in the year the heavy-footed adult Bloody-nosed Beetle (*Timarcha tenebricosa*) may be found lumbering its way amongst the grasses, and often upon the golden-flowered Gorse where it finds its mate (Fig. 8:12). When disturbed, this strange beetle produces a globule of red fluid from its mouth parts, repelling would-be predators. Its fat blue-black larva may be found clinging to grasses on which it feeds, together with the leaves of Bedstraws.

The world of the common in summer seems to change so quickly once the sun is obscured behind a cloud. The activities of many insects depend very largely upon the weather and time of the day. Try sweeping the same area at different times of the day and under different weather conditions. Always record your observations accurately. Just listen to this extract by W. H. Hudson:

Suction

Insect sucked in

8:13 'Pooter' used for collecting small insects

8:14 Can you spot the Grayling Butterfly?

8:15 A castle of foam (Cuckoo-spit) made by the soft-bodied young Frog-hopper and shown on the right devoid of its protective bubbles

Now most butterflies when they go to rest tumble anyhow into bed; in other words, they creep or drop into the herbage, take hold of a stem, and go to sleep in any position. . . . The blue has quite a different habit. . . . Instead of creeping down into the grass, they settle on the very tips of the dry bents. At some spots in an area of a few square yards they may be found in scores; one or two or three, and sometimes as many as half a dozen, on one bent, sitting head down, the closed wings appearing like a sharp pointed grey leaflet at the end of the stem. . . . When touched they scarcely move, and they will even suffer you to pick them off and replace them on the bent without flying away. . . .

Frog-hoppers and 'cuckoo-spit'

From late May onwards, the grass and flower stems become covered with mysterious blobs of white froth (Fig. 8:15). Gently part one of these frothy nests and you will find, snug inside, the squat-faced nymph of a Frog-hopper, with its blunt proboscis buried in the plant's cells. This tiny bubble maker can produce a sticky secretion through which air is blown by special tubes on its body, making it into a castle of foam, more persistent than any known detergent from a bottle! The nest of bubbles acts as protection against the nymph drying out and possibly also against enemies.

78

Hunting spiders and harvestmen

Here amongst the tangled grass stems, hundreds, perhaps thousands of spiders will be patrolling the ground, stealthily hunting their prey. Unlike other spiders, the *Lycosa* and *Pisaura* Wolf Spiders, as well as being cunning hunters, are also good mothers. They may be seen trundling pea-sized egg sacs around with them until the young hatch out. The larger pale-coloured *Pisaura* attaches the egg sac to grass, weaving the blades into a protective tent, over which she stands guard until the young spiderlings hatch (Fig. 8:17). The young of the *Lycosa* Wolf Spider, however, as soon as they hatch scramble on to her back and are carried around until they are too big.

In certain sandy areas even the Wolf Spiders have to watch out, for they are the prey of the spider-hunting Digger Wasp!

Ant hills

The Yellow Meadow Ant (*Lasius flavus*) produces the grass-covered mounds on many commons. The bulk of the ant population are wingless workers, spending their whole lives in various specialised duties such as foraging for food, caring for the eggs and larvae, or acting as guards for the colony. Notice in Fig. 8:16 how there is a steep side and a more sloping side to each mound. The ants inhabit the summit of the steepest slope, which invariably faces south-east to catch the morning sun's rays. The longest slope faces north-west and is caused by the progressive building and abandonment of new parts of the mound as the ant colony grows.

A press with the foot will tell you whether or not the mound is occupied. If soft and spongy it is occupied, but if it feels hard and unyielding it has been vacated by the ants – or perhaps a Green Woodpecker has worked at it to eat both larvae and adults.

Quite often ant mounds provide different habitats from the surrounding common. The plants which grow on them may be quite different from those on the adjacent soil for the formic acid given off by the ants will affect the soil of the mound. A study of these mounds will be well worth while.

It will be during the summer holidays that the swarming of the young male and female ants will take place. On a certain day a large number of the young winged ants leave the nesting

8:16 Nest hillocks of the Yellow Meadow Ant

8:17 Female Pisaura Hunting Spider stands guard over its egg cocoon

mounds over a large area, ascend the grass stalks and take off into the air. Pairing takes place in the air and the pair fall to the ground. There the male dies and the larger young queen bites off her wings and quickly seeks out a new nesting site.

However, you don't have to be on a common to witness this – quite often the black Garden Ants (*Lasius niger*), whose nests are sometimes concealed beneath paving stones in streets and gardens, suddenly erupt and the air becomes full of flying ants on their marriage flights.

'Save the village pond'

Most commons have sources of water – sometimes a small pool, perhaps a series of drainage ditches, or even a small stream. Many of these sources of water have become overgrown, some have been drained or filled in, since no longer are they needed to water the villagers' stock whilst grazing on the common. They are part of our heritage and form links with the history of the village, its church, its inn and common or green.

Discovery Assignment 8

1 Find out about your own local common and the part it has played in the history of your parish.
2 Make a study of grasses in your neighbourhood. Compare those found on a roadside verge with those at the edge of your playing field, or on a common.
3 Find out about the life cycles of Short-Horned Grasshoppers and the Long-Horned Bush Crickets.
4 Make a survey of 'cuckoo-spit'. Are any special plants chosen? Have any plants more bubble nests than others? At what height do the blobs occur?

9-WATER – FROM RAINWATER BUTTS TO CANALS AND STREAMS

Fresh water is perhaps one of the most interesting habitats to explore, for most of the major groups of animals and plants are represented in the life of ponds, lakes and streams.

Ponds in history

Prehistoric man probably relied upon streams, springs and natural ponds for his water supply. Later on, possibly the Iron Age settlers may have considered ways of collecting water. It would have been in the drier parts of the country where ponds were required. So that the water wouldn't run away the holes were lined with clay. This is probably the origin of the dewpond. The beautiful round dewpond at Ashmore, Dorset may go back to Anglo-Saxon or even Roman times.

Fish ponds were dug after the Norman invasion and stocked with fish for food. Some ponds occur where 'marl', a kind of chalky clay, was dug.

Many ponds, however, must have been made during the period when fields were being enclosed by hedges (Chapter 7).

A puddle and a microscope

Where water collects in tractor tracks (Fig. 9:2), particularly if the weather is warm, it quickly becomes green. This is due to masses of minute plants called algae. Look at a drop of this water beneath a microscope and you will discover all kinds of tiny plants and animals swimming around in the water. When a

9:1 The village pond is a unique habitat for aquatic creatures

9:2 Water-filled tractor tracks are quickly colonised by microscopic plants and animals

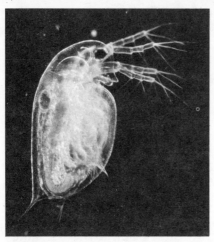

9:3 Micro-photograph of a Water-flea

9:4 The larvae and pupae of Gnats are often found in rainwater butts

puddle of water gets dried up many of these microscopic creatures produce a protective coat around themselves. These encysted forms are blown about by the wind, or are carried about on the feet of water birds, to colonise some other patch of water.

A rainwater butt and a hand lens

An open rainwater butt would not dry up as quickly as the water in the tractor track, so the water becomes greener and greener, with an ever-increasing population of a single celled animal-like plant – *Euglena,* and *Volvox. Volvox* is a sphere-like colony of plant cells, revolving slowly like a capsule in space. Tiny animals appear, just large enough to be seen with the naked eye. Using a hand lens you would find that there are two kinds moving jerkily about in the water – the flattened Water-flea (*Daphnia*) (Fig. 9:3) and the pear-shaped Copepod (*Cyclops*). These feed upon the plant cells.

In summer you will next notice larger creatures, each suspended head downwards from the water surface by a longish tube acting as a snorkel. This tube just pierces the surface film. By means of vibrating brushes on their heads these gnat larvae collect up the organisms on which they feed. Within two or three weeks the larvae change to a pupa looking very like a 'comma', not feeding at all but taking air through a pair of tiny 'ear trumpets'.

A pond and a hand net

Wherever a pond is made, if we are patient it will provide us with a complete community of life to explore. By simple observation and the careful use of a hand net we will be able to study some of the life on and beneath the surface. Lie quietly by the pond side and peer through the clear water. You should see the patterns of life being enacted by water creatures that hunt, and others that are the hunted.

Imagine you are but a few millimetres tall and have visited the bottom of the pond in a miniature bathysphere. The creatures you collect from the submerged plants and open water will take on a new fascination, especially when viewed through a hand lens. The larger the creature the fewer will be found. The simplest food chain begins, as we saw earlier, with microscopic plants which are eaten by microscopic animals. These in turn provide food for small animals and insects. Larger insects, such as Water Boatmen, feed upon the smaller ones, and themselves may be gobbled up by the ferocious *Dytiscus* Water Beetle, or by fish, or water birds if they are present.

Dwellers everywhere

All parts of the pond are occupied. The water surface is like a thick skin on which a number of creatures live and feed. The Water Skater (*Gerris*) (Fig. 9:5) strides over the pond water, making use of the surface tension, pouncing on any hapless fly which falls into the water. Whirligig Beetles (*Gyrinus*) swim erratically in tiny circles on the surface as they hunt for flies and small insects.

The Water Scorpion (*Nepa cinerea*) sits motionless on water

9:5 The carnivorous Water Skater seen from above

9:6 The ferocious larva of the Great Diving Beetle will tackle most creatures living in the pond

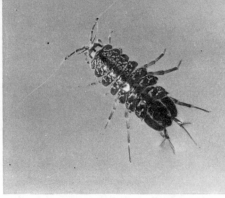

9:7 The Notonecta Water Boatman is a water-bug. It has a sharp piercing 'beak' used for sucking its prey

9:8 The Water Slater spends the whole of its life in the water

plants in shallow water, its bristle-like breathing tube reaching up to the surface, waiting to shoot out its scissor-like forelegs as a water insect passes. Amongst the weeds in deeper water the baleful nymphs of Dragonflies creep stealthily, attacking and eating creatures far larger than themselves. They catch their prey by flicking forward a set of false jaws which act like a double-hinged excavator boom, folding back beneath the head when not in use.

In the open water, Water Boatmen (*Notonecta glauca*) (Fig. 9:7) row strongly upside-down. Its near relative, *Corixa*, however, swims more weakly.

The Water Slater (*Asellus aquaticus*) (Fig. 9:8) creeps along the bottom, feeding upon decomposing leaves or other material.

9:9 Some watery life cycles

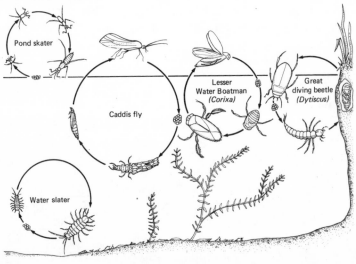

Some watery life cycles

There are two main types of aquatic life. Pond Snails, Water Shrimps and Water Slaters are true residents, enacting their complete life cycle beneath the water (Fig. 9:9). Water Beetles, Water Bugs, Caddis Flies, May Flies and Dragonflies only spend part of their life cycle in the water, showing adaptations to this life in widely different ways.

Feeding habits and the aquaria

Some aquatic creatures cling so tightly to submerged vegetation that they are seldom caught with a net. Put some of this vegetation into a shallow container, such as a washing-up bowl, together with enough pond water to cover it and allow it to settle. Within a short while you will be able to observe many of the inmates moving around and perhaps feeding. The large carnivores should be removed to separate containers or you will find that overnight many of the herbivores, or decomposers, will have disappeared.

Water animals, if properly cared for and fed, will live satisfactorily in an aquarium or, depending upon their size, in smaller containers such as large sandwich boxes or wide-mouthed jars. Make sure the surface area of the container is large enough to give the animals sufficient oxygen. Do *not* use water straight from the tap since the chlorine it contains may kill your animals. Either use rain water or allow tap water to stand out of doors for a few days.

Plant life of watery habitats

Liverworts, we have discovered, are to be found in moist habitats, and mosses and ferns require shady conditions at certain times of their life cycles. In Fig. 9:10 is an example of a very

9:10 Unusual water plants. Floating Liverwort (*Riccia*) amongst the larger common duckweed and frogbit

9:11 Floating Fern (*Azolla*)

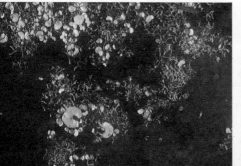

9:12 Even the canal flowing through an industrial city will provide a number of plant and animal communities

9:13 Nature never stands still. See how this disused canal has been choked by the rapidly growing *Phragmites*

primitive Liverwort (*Riccia fluitans*) which grows in floating masses in water containing a fairly high content of dissolved minerals.

In the same ornamental pond there suddenly appeared large numbers of the tiny floating Fern (*Azolla filiculoides*) (Fig. 9:11), whose fronds become tinged with red in the autumn. It is not a native plant but has been introduced from the warmer parts of America.

Slow-flowing streams and brooks are the places to look for the largest moss growing in Britain. The Willow Moss (*Fontinalis antipyretica*) grows attached to submerged logs or stones, but seldom produces spore capsules unless the water level drops, leaving the plant out of water.

The briefest of looks at the vegetation around a pond or canal should reveal the presence of several distinct communities of flowering plants adapted to the conditions in which they grow (Fig. 9:12). Some plants, like the Water Crowfoot (*Ranunculus sp.*) may cover the whole surface of the water with their showy white flowers, whereas the flowers of the Water Starwort (*Callitriche sp.*) are hardly noticeable. Many water plants produce very little seed, relying on fragments of plants becoming detached and rooting elsewhere. This is called vegetative reproduction.

Canals and their life

The British canal system provides a whole range of water habitats, very similar in some respects to a large pond. From the towpath it is possible to study many parts of the canal habitat, its plants and animals, including the herbivorous bank-living Water Vole, the brilliant Kingfisher skimming the water surface, and the Mute Swans that nest amongst the reeds which have colonised an unused part of the 'cut'.

Here, in canals, you can see how dynamic are the forces of Nature. Unless they are periodically cleared, the canal waterways rapidly become choked with encroaching swamp vegetation, providing, as in Fig. 9:13 further habitats for yet other creatures. Many a canal passes through built-up areas and unfortunately may be subject to the effects of industrial waste, which will destroy much of the water life therein (Fig. 9:14).

Streams and rivers – living indicators of their cleanliness

Freshwater animals, particularly those living in streams and rivers, vary in the amount of oxygen they require. The amount of oxygen dissolved in flowing water has an important effect upon their distribution. We can use this principle to find out whether or not a stream is polluted.

The larvae of the *Chironomus* midge and small worms called *Tubifex* have a red pigment (haemoglobin) in their blood, enabling them to live in very stagnant conditions where there is very little available oxygen. Flatworms and small crustaceans such as the Water Slater (*Asellus*) have a low oxygen requirement, absorbing all they need through their skin. The larvae of Caddis Flies and the nymphs of May Flies have specially developed breathing organs (gills) and require more oxygen. Others, such as the Stoneflies, require a lot of oxygen to live, and obviously are the first to suffer if, by some form of pollutant, the natural oxygen content of the stream is lowered.

By the presence or absence of some, or all, of these different creatures in a stretch of flowing water we can get some idea as to how clean, or polluted, it is.

9:14 Waste chemicals released into the water by canalside factories killed these fish

9:15 Contaminated fish would kill the brilliant Kingfisher, for it is at the end of a canal food chain

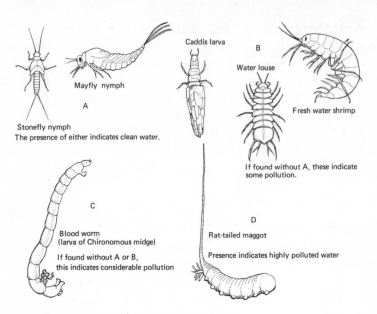

Caddis larva

Mayfly nymph

A

B

Water louse

Fresh water shrimp

Stonefly nymph
The presence of either indicates clean water.

If found without A, these indicate
some pollution.

C

Blood worm
(larva of Chironomous midge)

If found without A or B,
this indicates considerable pollution

D

Rat-tailed maggot

Presence indicates highly polluted water

9:16   Living indicators of a stream's cleanliness

Looking around

Although it would be wonderful to have ready access to a canal, a flooded gravel pit or some similar area, be prepared to take advantage of any watery habitat that you come across. Quite often there are wet marshy areas, or even temporary pools on waste ground, by rubbish tips and on building sites. On these unknown areas *do be very careful*.

Discovery Assignment 9

1 Using old maps, discover all you can about the pools, canals and water courses of your area. How many are in existence today? Are there any worth saving?

2 Stand two jam jars of rain water outside, in which one has about a centimetre of garden soil. Compare the animal life which appears.

3 Make a study of the fish of canals and rivers. Find out what they feed upon and the conditions they require for spawning.

4 Discover and illustrate the various ways by which aquatic animals breathe, move or feed.

5 If you are able to use a small microscope observe the animal life living on a fragment of water weed, a piece of decaying leaf, the green 'slime' from a stagnant puddle.

10-FURTHER STUDIES AT HOME

You may have encountered many creatures strange to you during your expeditions into the unknown worlds of the brick wall, the vacant building plot, or the compost heap at the bottom of your garden. You will perhaps wish to look more closely at the lives of some by creating temporary habitats on your window-sill. Here are a few ideas:

Window boxes and plant tubs

Perhaps you have no back garden. Would it be possible for you to have a window box, or one of the many kinds of plant tubs now available? These could be planted up with flowers such as Sweet Alyssum, Stocks and Tobacco Plants, which are especially attractive to insects. Record the flower visitors – how long do they stay? – does one colour attract them more than another? (e.g. purple Alyssum is better than white). How long is your portable garden free from weeds? What arrives first? How long is it before aphids colonise your plants?

Moth traps – light, and sugar

There is much we have to learn about the moth populations of our towns and cities. It is not necessary to have a garden either, for moth traps can be used on a flat roof, like the one used at Gloucester City Museum. The special mercury-vapour light traps are expensive but you could construct a simple one using an ordinary 150 watt tungsten bulb. A portable camping lamp standing on a white sheet, on which are some egg trays, will attract quite a large number of moths in summer.

Many naturalists in past years used a technique called 'sugaring'. The ingredients were much cheaper then than now. However, black treacle painted on to trees or posts in your garden, or on to rolls of corrugated cardboard tied to trees and posts outside your garden, will attract numerous insects. This is best done at dusk, on close warm nights, with a good cover of cloud in the sky.

A home for ants – the formicarium

There are a number of ways shown in various books of keeping ants. One of the simplest is made of two glass sheets with a wall

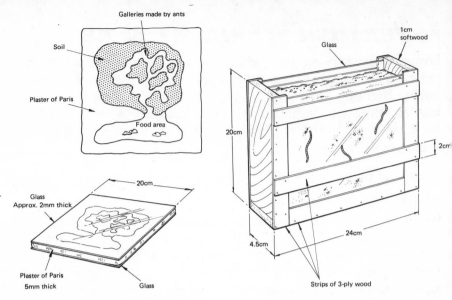

10:1 An easily made Formicarium for observing ants

10:2 This Wormery can be adapted to house Dung Beetles

of plaster of Paris between (Fig. 10:1). The reason why plaster is used is that it can be moistened from time to time – ants quickly die if they have no access to water. Put food, such as a few drops of honey or concentrated sugar solution into one section and the ants and soil into the other.

When digging the ants out of the nest include eggs, larva and pupa and if possible the larger queen. Sometimes you may come across very small white woodlice living with the ants. These could be included as well.

Remember to cover the glass with black paper or card – ants will not construct their galleries in the light.

The wormery

Worms, like ants, can be kept for short periods in glass jars of soil, covered round with black or brown paper. However, the construction shown in Fig. 10:2 enables you to observe the activities of the worms much better. There is no need for a more permanent construction unless you are capable of using wood-working tools.

Make sure the soil is not acid. Worms need a certain amount of calcium available to them. They will also need food, such as rotting leaves, placed on the soil surface.

Keeping Dung Beetles

Dung Beetles, such as the 'Lousy Dor', or Minotaur, may be kept in soil with a little moist dung, inside large perspex sandwich boxes. It is possible to watch them laying eggs if you put them into an 'elongated wormery' of glass. The top of the soil should be capped with either cow or rabbit dung according to the species and, of course, where it was found.

Dung Beetles are very important creatures of pasture land consuming as they do vast quantities of cow pats throughout the year. The study of a cow pat, too, provides a wonderful example of animal succession.

Rearing butterflies and moths

Many books are available on the subject of rearing butterflies and moths, both from the egg and larval stages. It is not necessary to have complicated cages – perspex boxes in a variety of sizes, with some butter muslin and rubber bands, will enable you to observe the larvae in small numbers.

So little has been written in detail about the life histories of many species, let alone other kinds of insects. As you know, at intervals during the caterpillar's life it stops eating for a day or so. During this period a new skin is growing beneath the old one.

10:3 Perspex boxes of various sizes will keep the larval food plant fresh

10:4 Simple cylinder-type cage, made from a tin, its lid and an acetate sheet

When formed, the old skin splits behind the head and the larva creeps out, leaving the old skin behind. This is called moulting, or 'ecdysis', the stages between moults being known as 'instars'. Some larvae eat the old skin, others leave it where it was discarded. Many larvae change colour as they grow. All this should be recorded. You may have to await the emergence of the insect from its pupa before you are able to name it. If you cannot name it from a reference book, take it to your local museum for identification.

Record, too, how much is eaten during each instar (Fig. 10:5).

A zoo on your window-sill

You can really only learn about the inter-dependence between plants and animals by observing at first hand in the field (even though that might be the street on the way to school). By doing this you will be able to call yourself a field naturalist. However, the habits of many freshwater inhabitants will have to be studied

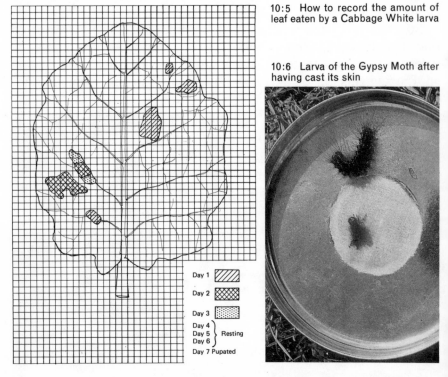

10:5 How to record the amount of leaf eaten by a Cabbage White larva

10:6 Larva of the Gypsy Moth after having cast its skin

Day 1
Day 2
Day 3
Day 4
Day 5 } Resting
Day 6
Day 7 Pupated

at home or at school. This is where a jam-jar zoo is a valuable possession, provided the jars are kept in a shady place.

Water Snails, Water Boatmen, Caddis Fly larvae, May Fly nymphs and Dragonfly nymphs may all be kept provided they are well fed. This will mean supplies both of waterweed and of the smaller crustaceans (*Daphnia* and *Cyclops*) as food.

A home museum

As a field naturalist, in studying different habitats and their communities, you will be collecting records of observations. A museum is a place of reference, using both books and preserved specimens. However, on your expeditions you will come across various evidences of animal activity, such as galls or leaf miners. These, if mounted carefully, will be very useful in supplementing your records, drawings and photographs. The food remains found on spiders' webs, and the cast skins of insects are also useful.

Codes of Conduct

Although much of this book has been about habitats which could be found in a town or urban situation, you should always go about your investigations as if you were working in a much more fragile habitat in the open countryside. This is good training and will win you respect from the many people who will be wondering what you are doing in those out-of-the-way places of towns and cities. A list of the available Codes of Conduct appears at the end of the book.

Conservation and you

Much of this book has been about the diversity of life in the most commonplace of habitats. Perhaps by the time you have reached this page you may have even discovered for yourself some of the jigsaw pieces of wildlife in garden, park or neglected open space? If in these places there is a multitude of life, how much more will there be in the wilder parts of the countryside?

In the dictionary the word 'conserve' is defined as 'keep from harm, decay or loss'. You may have noticed articles in various

10:7 Conservation in practice – Nature Reserve in grounds of Burford County Primary School, Oxfordshire

newspapers, or television news items, about the decline of the Otter, the gassing of Badgers, the protection given to the Ospreys of Scotland, and to the Red Kites of Wales. Far less attention, though, is given to the smaller animals of the world. If you feel that the smaller animals must be protected, then you can help by talking about them to your friends and your parents.

There is little doubt that our towns and cities will keep creeping slowly outwards, spreading fingers of concrete over the good soil. If we all learn something about the lives of the plants and animals around us there is no doubt we will be in a better position to care for them, and to protect their habitats in the future.

Discovery Assignment 10

1 Design a pictorial code of conduct which could be used as a guide to friends or pupils of another school, when investigating town habitats.
2 Make a scrapbook of cuttings about Nature Conservation in (*a*) local newspapers and (*b*) in national newspapers.
3 Keep a diary of observations made throughout the seasons on either a town or a country habitat.
4 After having observed the plants and animal life of a piece of neglected ground you hear that the local council are going to bulldoze the area clear. Compose a letter to them telling them your ideas of making it into an urban nature reserve.
5 Find out about the work being done by your own County Naturalists Trust. Write a letter to them asking how you can help.

Books For Further Reading

Bainbridge J. *Conservation* Evans
Darlington A. *Pollution and Life* Blandford
Duncan V. *Animals and Plants in the Fields* A. & C. Black
Finch I. *Animals in the Soil* Longman
Finch I. *Pond Animals* Longman
Jackman L. A. *Exploring the Hedgerow* Evans
Leutscher A. *The Ecology of Water Life* Franklin Watts
Mabey, R. *The Pollution Handbook* Penguin Education
Perry G. A. *Soils* (Rural Studies Series) Blandford
Perry G. A. *Plant Life* (Rural Studies Series) Blandford
Reynolds C. *Small Creatures in my Back Garden* Target Books
Richards J. *The Hidden Country* Faber
Simmons G. E. *Field Studies* Longman
Watson G. G. *Fun with Ecology* Kaye and Ward
Williams C. *What you can find in a Park* A. & C. Black

Useful Addresses

Amateur Entomologists' Society, 16 Frimley Court, Sidcup Hill, Sidcup, Kent.
British Museum (Natural History), Cromwell Road, London SW7.
Council for Nature, c/o Zoological Society, Regent's Park, London NW1 4RY.
Monkswood Experimental Station, Abbots Ripton, St Neots, Huntingdonshire.
Royal Society for the Protection of Birds, The Lodge, Sandy, Bedfordshire.
Society for the Promotion of Nature Reserves, The Green, Nettleham, Lincoln (for addresses of County Naturalists Trusts).

Codes of Conduct

Wild Flowers. Botanical Society of the British Isles, c/o British Museum (Natural History).
Insects. Royal Entomological Society of London, 41 Queens Gate, London SW7.
Wildlife Photography. Royal Society for Protection of Birds.

INDEX